Facilitator's Guidebook
for Use of Mathematics Situations
in Professional Learning

Facilitator's Guidebook
for Use of Mathematics Situations
in Professional Learning

Edited by

Rose Mary Zbiek
Glendon W. Blume
M. Kathleen Heid

Copublished by:
NCTM, NCSM
&

INFORMATION AGE PUBLISHING, INC.
Charlotte, NC • www.infoagepub.com

Library of Congress Cataloging-in-Publication Data

The CIP data for this book can be found on the Library of Congress website (loc.gov).

Paperback: 978-1-64113-079-0
Hardcover: 978-1-64113-080-6
eBook: 978-1-64113-081-3

NCSM is a mathematics education leadership organization that equips and empowers a diverse education community to engage in leadership that supports, sustains, and inspires high quality mathematics teaching and learning every day for each and every learner

Printed in the United States of America

CONTENTS

PREFACE

Mathematics matters. It especially matters for teachers of mathematics and for those who work with teachers in in-service and pre-service settings. The purpose of this Guidebook is to be a tool for facilitators who help teachers to make connections among secondary school mathematics ideas and to value the relevance and power of their understanding of mathematics beyond that which they teach.

The Guidebook and the six Situation Guides it contains are for use in professional learning. Although the volume can be used by an individual teacher, the materials are designed for a facilitator to use with small groups or large groups of teachers. The group might be in a school setting, in a pre-service mathematics or mathematics education course, or in another venue in which teachers are working to enrich their understandings of mathematics and to improve their practice.

One unique feature of the Situations in this book, and the larger set of Situations in *Mathematical Understanding for Secondary Teaching: A Framework and Classroom-Based Situations* (Heid & Wilson, 2015), is their origin in the work of teachers.

Facilitator's Guidebook for Use of Mathematics Situations in Professional Learning,
pages vii–viii.
Copyright © 2018 by Information Age Publishing
All rights of reproduction in any form reserved.

Rather than imagining the mathematics that teachers encounter, the Situations Project team drew from incidents they had witnessed in the daily work of teaching mathematics. These incidents become prompts to engage teachers in mathematics.

This book is for facilitators and for teachers who seek new mathematical experiences that reveal secondary mathematics as a connected and consistent subject that teachers and their students can enjoy.

ACKNOWLEDGEMENTS

The Situations Project (Principal Investigators and Senior Research Associates: M. Kathleen Heid, Patricia S. Wilson, James W. Wilson, Glendon Blume, Jeremy Kilpatrick, Rose Mary Zbiek) was supported in part by the National Science Foundation under Grant ESI-0426253 for the Mid-Atlantic Center for Mathematics Teaching and Learning (MAC-MTL) and Grant ESI-0227586 for the Center for Proficiency in Mathematics Teaching (CPTM). This publication arose from a collaboration of the National Council of Supervisors of Mathematics and the Situations Project Principal Investigators and Senior Research Associates. Any suggestions, opinions, findings, or conclusions expressed in this document are those of the authors and do not necessarily reflect the views of the National Science Foundation.

The MAC–CPTM Situations Project

M. Kathleen Heid, Distinguished Professor
The Pennsylvania State University
271 Chambers Building
University Park, PA 16803
(814) 865-2226
mkh2@psu.edu

Patricia S. Wilson, Professor Emerita
University of Georgia
Department of Mathematics and Science Education
105H Aderhold
Athens, GA 30602
(706) 542-4547
pswilson@uga.edu

The National Council of Supervisors of Mathematics

2851 S. Parker Rd. #1210
Aurora, Colorado 80014
(303) 317-6595
office@mathedleadership.org
mathedleadership.org

CHAPTER 1

INTRODUCTION TO FACILITATING
THE USE OF SITUATIONS

Rose Mary Zbiek

During a typical school day, mathematics teachers encounter incidents that call on their mathematical understandings in the course of the work of planning and implementing mathematics lessons. The contexts of the incidents vary and include such things as a student asking a question, a surprising statement in curriculum materials, and a conjecture posed by a group of students. Regardless of the context, the ways in which teachers understand mathematics and use that understanding when these incidents occur distinguish the responsibilities of mathematics teachers from those of other professionals who use mathematics in their daily work. This book provides guidance in how to use these incidents to engage teachers in learning and applying mathematics to enrich their professional practice.

Facilitator's Guidebook for Use of Mathematics Situations in Professional Learning,
pages 1–5.
Copyright © 2018 by Information Age Publishing

THE SITUATIONS PROJECT

A given incident can evoke a range of mathematical understandings, but not all of those understandings are immediately available to teachers. Recognizing the need to understand what mathematics secondary teachers can productively use, and inspired by their own teaching experience and enjoyment of mathematics, mathematics educators at The Pennsylvania State University and the University of Georgia collected descriptions of incidents that they had witnessed in the work of teaching secondary school mathematics. The educators then developed descriptions of mathematical ideas that a teacher might productively use in response to each incident.

Each of these combinations of descriptions of incidents and related mathematical ideas constitutes what their creators call a *Situation*. Each Situation consists of a *Prompt* (a description of the incident), several *Mathematical Foci* (descriptions of mathematical ideas pertinent to the incident), and a *Commentary* (a discussion of overarching ideas in and connections among the Mathematical Foci). Funded in part by the National Science Foundation under Grants ESI-0426253 and ESI-0227586, the Situations Project team created, refined, and revised more than 50 of these Situations that address algebra, number, geometry, and statistics. The collection of Situations then achieved two purposes.

To achieve one purpose, an in-depth analysis of these Situations informed the group's creation of a framework for Mathematical Understanding for Secondary Teaching (MUST). The MUST framework, elaborated by Kilpatrick, Blume, Heid, J. Wilson, P. Wilson, and Zbiek (2015), identifies and elaborates the mathematical understanding that secondary mathematics teachers could productively use. This understanding is viewed from three perspectives: (a) mathematical proficiency, (b) mathematical activity, and (c) the mathematical context in which teachers can productively draw on mathematical understandings.

The perspective of *Mathematical Proficiency*, which includes such things as conceptual understanding and strategic competence, extends and enriches students' mathematical proficiencies that are articulated in *Adding It Up* (National Research Council, 2001). In particular, the framework authors characterize mathematical proficiency for secondary teaching in a way that draws on the five existing strands found in *Adding It Up* and supplements them with a sixth strand, Historical and Cultural Knowledge.

The team identified the perspective of *Mathematical Activity* as actions involved in doing mathematics. The naming of strands in this perspective focuses on neither procedures nor algorithms but rather on such actions as Mathematical Noticing and Mathematical Creating that capture how teachers, students, and others engage with mathematics. Explicit mention of mathematical activity acknowledges mathematics as a dynamic, creative, and meaningful human activity.

The strands of the perspective of *Mathematical Contexts of Teaching*, such as Know and Use the Curriculum and Access the Mathematical Knowledge of Learners, recognize the settings in practice in which teachers draw on mathematical proficiency and engage in mathematical activities. The contexts are not necessarily venues in which other professionals use mathematics.

The second purpose of the Situations centers on their usefulness in engaging teachers of all experience levels in foregrounding mathematical ideas, deepening understandings, and connecting various parts of their mathematical experiences to school mathematics. Each Situation delves deeply into mathematical ideas on which school mathematics focuses and builds. The Situations creators worked from incidents in their own work as teacher educators, mathematics instructors, and professional development providers. The Situations also drew the attention of others, including National Council of Supervisors of Mathematics (NCSM) leaders and groups.

This book arises from collaborations between a team of NCSM leaders and Situations Project faculty personnel from The Pennsylvania State University and University of Georgia. Its purpose is to provide six explicit examples of Situations and how they might provide the basis for professional learning. The six examples might inspire ideas about how other Situations might be used with in-service and pre-service teachers.

THE SIX SITUATIONS

The six Situations selected from Heid and Wilson (2015) for inclusion in this book resonate with the experience of many teachers and can arise in different secondary school mathematics curricula and courses. Considered collectively, they represent a range of mathematics content across algebra, number, geometry, and statistics. Some Situations address recurrent topics that arise in many places across middle grades and high school curricula. For example, Product of Two Negative Numbers might arise with the factoring of whole numbers in middle school and with the factoring of polynomials in algebra settings.

In addition to having potential for work with secondary teachers, some of the topics have potential for use in professional development settings or for work with teachers who teach or will teach elementary mathematics. For example, Division Involving Zero is a phenomenon that arises in children's work with multiplication and division in Grades 3–5, in work with rational expressions in algebra, and again with limits in calculus.

Taken individually, some of the Situations can advance multiple professional learning goals. For example, Mean and Median offers the opportunity to help teachers further their understanding of descriptive statistics. This Situation also is useful in making the overarching point that two different types of representations of any one statistical or mathematical concept or relationship might efficiently convey very different information about the topic at hand.

USING THIS GUIDEBOOK

The purpose of this Facilitator's Guidebook is to provide both several examples of Situations and ideas for how professional developers or teacher educators might use such Situations in sessions with prospective or practicing teachers. The Situations were developed from teaching dilemmas in secondary school mathematics classrooms, and these dilemmas are faced by student teachers as well as by their veteran colleagues. The Situations offer unique opportunities for teachers to engage in mathematics that is directly related to the secondary school curriculum. The Situations are a venue for teachers to use mathematical knowledge and to develop their own deeper understanding of secondary school mathematics.

The Foci and Commentaries connect these ideas to each other and to ideas that secondary mathematics teachers encounter in college mathematics courses. In some Situations, Foci and Commentaries highlight content that students first encounter in their elementary school years. The Situations are a tool to help teachers further develop understandings of mathematics that include and extend ideas typical of secondary school curricula as they underscore mathematics as a coherent and connected body of knowledge.

USE OF THIS BOOK

This Guidebook is offered as a tool for facilitators of professional learning for in-service and pre-service teachers of mathematics. The book might be used in part or in whole with any one group, and single Situations might be the focus of a professional learning session. For example, one Situation might be used for professional learning within a school that is focused on what it means to have a coherent and connected curriculum. Another Situation might be used as a tool to encourage mathematical conversation among teachers of different mathematics topics or courses or among teachers who work primarily at different grade levels.

Taken in its entirety, the book can be used in various ways to engage teachers in mathematics. For example, it might be used in an academic-year-long sequence of sessions, during an intensive summer experience, or within a mathematics content course for teachers. In addition, teachers and facilitators might choose to identify incidents in their own practice and use them to create and share original Situations based on their own teaching practice. Those who engage in this creative work might find *Mathematical Understanding for Secondary Teaching: A Framework and Classroom-Based Situations* (Heid & Wilson, 2015) to be helpful. In particular, chapter 5 of that volume, Creating Situations as Inquiry (Zbiek & Blume, 2015), offers insights into ways that developing Situations might serve as professional learning work that enhances practice. In addition, this volume contains another 37 Situations that could be used in professional learning settings.

STRUCTURE OF THIS BOOK

This Guidebook begins with this introduction to the book, an explanation of what Situations are, and ideas about their use with experienced and emerging teachers. The volume ends with a chapter that brings closure to the examples and offers general thoughts about using Situations in professional learning situations.

Chapters 2 through 7 are the heart of the book. Each of these chapters focuses on one Situation and how teachers might interact with it. Each chapter presents an outline of how teachers might engage with the content focus of the Situation and includes both reflection questions and a copy of the Situation. Suggestions offered for implementing each Situation can be adapted for use with many different teacher-learning communities and in various settings.

CONCLUSION

Mathematics is a rich and glorious field of connected ideas. Secondary school mathematics involves many topics that connect with ideas encountered at various places in school and college curricula. This Guidebook provides ideas and plans for how facilitators might capitalize on existing Situations in their work with teachers to promote, connect, and enrich understandings of secondary school mathematics. We hope that this book helps you and others to engage teachers in mathematics; to inspire them to probe ideas and take mathematical chances; and to delight with them in the coherence and consistency of secondary school mathematics.

REFERENCES

Heid, M. K., & Wilson, P. S. (with Blume, G. W.) (Eds.). (2015). *Mathematical understanding for secondary teaching: A framework and classroom-based situations.* Charlotte, NC: Information Age.

Kilpatrick, J., Blume, G., Heid, M. K., Wilson, J., Wilson, P., & Zbiek, R. M. (2015). Mathematical understanding for secondary teaching. In M. K. Heid & P. S. Wilson (with G. W. Blume) (Eds.), *Mathematical understanding for secondary teaching: A framework and classroom-based situations* (pp. 9–30). Charlotte, NC: Information Age.

National Research Council. (2001). *Adding it up: Helping children learn mathematics.* J. Kilpatrick, J. Swafford, & B. Findell (Eds.). Mathematics Learning Study Committee, Center for Education, Division of Behavioral and Social Sciences and Education. Washington, DC: National Academy Press.

Zbiek, R. M., & Blume, G. (2015). Creating new situations as inquiry. In M. K. Heid & P. S. Wilson (with G. W. Blume) (Eds.), *Mathematical understanding for secondary teaching: A framework and classroom-based situations* (pp. 57–63). Charlotte, NC: Information Age.

CHAPTER 2

FACILITATOR'S GUIDE
FOR
DIVISION INVOLVING ZERO

Situation 1 From the MACMTL–CPTM Situations Project[1]

Diane Briars, Rose Mary Zbiek, Glendon Blume, M. Kathleen Heid,

M. Suzanne Mitchell, Connie Schrock, Steven S. Viktora, and James W. Wilson

Division involving 0 is one mathematical topic that resonates with the experience of many teachers. Every mathematics teacher

at one time or another is faced with the dilemma of interpreting expressions such as $\frac{n}{0}$ or $\frac{0}{0}$. Students may ask about it, text-

books may introduce the idea, or the development of a given mathematical idea may require dealing with interpretation of

such indicated calculations. This chapter offers ideas for exploring in a professional learning or teacher preparation context a

Situation about division involving 0.

Facilitator's Guidebook for Use of Mathematics Situations in Professional Learning,
pages 7–27.
Copyright © 2018 by Information Age Publishing

OVERVIEW

Facilitators can scan the following overview to quickly get a sense of the mathematics involved in the proposed professional learning setting. The sections deal with the mathematics of the Division Involving Zero Situation, why these mathematical ideas might be important for participants, the learning goals for participants, and mathematical ideas central to the proposed professional learning sessions.

Situation	Relevance
This Situation addresses the possible values that result when 0 is the dividend, the divisor, or both the dividend and the divisor in an indicated quotient, such as: • $0 \div n = ?, n \neq 0$ • $n \div 0 = ?, n \neq 0$ • $0 \div 0 = ?$ Division involving 0 can be viewed through multiple contexts (e.g., arithmetic operations, real world analogies, rates and ratios, and the real projective line).	• Preservice and in-service teachers at all levels have misconceptions about division involving 0, including that $0 \div 0 = 0$, or $0 \div 0 = 1$. They regularly are put in the position of needing to explain such division. • Even when teachers know correct answers for problems addressing division involving 0, their understanding may be limited. They may cite rules as the reason and may not be able to provide a valid mathematical explanation.
Goals	**Key Mathematical Ideas**
• Understand why dividing 0 by a nonzero number is 0 and division by 0 is undefined or indeterminate. • Distinguish results that are undefined from results that are indeterminate. • Clarify common misconceptions about division involving 0. • Consider how and when to address this issue with students.	• When 0 is the dividend only, the divisor only, or both the dividend and the divisor in a quotient, the value of such a quotient is 0, undefined, or indeterminate, respectively. • There is a distinction between undefined and indeterminate. • Division involving 0 can be connected to: factor pairs, Cartesian product, and area of rectangles (Focus 3); ratios and rates (Focus 4); and the real projective line (Focus 5).

The Situation under consideration in this chapter follows (highlighted with a gray background).[2]

DIVISION INVOLVING ZERO

Situation 1 From the MACMTL–CPTM
Situations Project

Bradford Findell, Evan McClintock, Glendon Blume,

Ryan Fox, Rose Mary Zbiek, and Brian Gleason

PROMPT

On the first day of class, pre-service middle school teachers were asked to evaluate $\frac{2}{0}, \frac{0}{0}$, and $\frac{0}{2}$ and to explain their answers. There was some disagreement among their answers for $\frac{0}{0}$ (potentially 0, 1, undefined, and impossible) and quite a bit of disagreement among their explanations:

- Because any number over 0 is undefined;

- Because you cannot divide by 0;

- Because 0 cannot be in the denominator;

- Because 0 divided by anything is 0; and

- Because a number divided by itself is 1.

COMMENTARY

The mathematical issue centers on the possible values that result when 0 is the dividend, the divisor, or both the dividend and the divisor in a quotient. The value of such a quotient would be 0, undefined, or indeterminate, respectively. The Foci use multiple contexts within and beyond mathematics to represent and illustrate these three possibilities. Connections are made to ratios, factor pairs, Cartesian product, area of rectangles, and the real projective line.

MATHEMATICAL FOCI

Mathematical Focus 1

An expression involving real number division can be viewed as real number multiplication, so an equation can be written that uses a variable to represent the number given by the quotient. The number of solutions for equations that are equivalent to that equation indicates whether the expression has one value, is undefined, or is indeterminate.

One can think of a rational number as being the solution to an equation. If division expressions involving 0 also represent rational numbers, equations involving these expressions should have consistent results. To find the solution of the equation $\frac{0}{2} = x$, consider the equivalent statement $2x = 0$, which yields the unique solution $x = 0$. To see the impossibility of a numerical value for a rational number with a 0 in the denominator, consider the equation $\frac{0}{0} = x$, and its potentially equivalent equation, $0x = 0$. Because any value of x is a solution to this equation, there are infinitely many solutions; hence, there is no unique solution, and so the expression $\frac{0}{0}$ is indeterminate. Using the same thinking, if $\frac{2}{0} = x$, then $0x = 2$. No real number x is a solution to this equation, so the expression $\frac{2}{0}$ is undefined.

Mathematical Focus 2

One can find the value of whole number division expressions by finding either the number of objects in a group (a partitive view of division) or the number of groups (a quotitive view of division).

In partitive division, a given total number of objects is divided equally among a number of groups. A nonzero example would be $\frac{12}{3}$, in which 12 objects are shared equally among 3 groups and the question concerns how many objects would be in one group. Similarly, $\frac{0}{2}$ can be thought of as 0 objects in 2 groups, which means 0 objects in each group. Additionally, the expression $\frac{0}{0}$ is a model for dividing 0 objects among 0 groups. In other words, if 0 objects are shared by 0 groups, how many objects are in 1 group? There is not enough information to answer this question, so the expression $\frac{0}{0}$ is indeterminate. If the number of objects in a group is 3, or 7.2, or any size at all, 0 groups would have 0 objects. Similarly, $\frac{2}{0}$ is a model for the question: If 2 objects are shared by 0 groups, how many objects are in 1 group? In this case the number of objects in the group is undefined, because there are 0 groups.

Using a quotitive view of division, the expression $\frac{12}{3}$ is interpreted as a model of splitting 12 objects into groups of 3 and asking how many groups can be made. So $\frac{0}{2}$ can be thought of as splitting 0 objects into groups of 2, which means 0 groups of size 2. The expression $\frac{0}{0}$ models the splitting of 0 objects into groups of size 0, and asks how many groups can be made. Because there could be any number of groups, there are an infinite number of solutions, so the expression is indeterminate. Lastly, the expression $\frac{2}{0}$ models the splitting of 2 objects into groups of 0 and asking how many groups can be made. Regardless of how many groups of 0 are removed, no objects are removed. Therefore, the number of groups is undefined.[1]

Mathematical Focus 3

The mathematical meaning of $\frac{a}{b}$ (for real numbers a and b and sometimes, but not always, with b ≠ 0) arises in several different mathematical settings, including slope of a line, direct proportion, Cartesian product, factor pairs, and area of rectangles. The meaning of $\frac{a}{b}$ for real numbers a and b should be consistent within any one mathematical setting.

There are mathematical situations in which ratios are necessary, and a quotient can be reinterpreted as a ratio. For example, the slope of a line between two points in the Cartesian plane can be defined as the ratio of the change in the y-direction to the change in the x-direction, or as the rise divided by run. In the case of two coincident points, the change in the y-direction and the change in the x-direction are both 0, which means that the rise divided by run is $\frac{0}{0}$. There are an infinite number of lines through two coincident points, and so the slope is indeterminate. In the case of two points lying on the same vertical line whose y-coordinates differ by a, the change in the y-direction is a, and the change in the x-direction is 0. It might be tempting to claim that because the slope of a vertical line is undefined, $\frac{a}{0}$ is undefined. However, this claim is exactly what needs to be shown.

The model for direct proportion, $y = kx$, represents a family of lines through the origin. For y and nonzero x as the coordinates of points on a line given by $y = kx$, the ratio $\frac{y}{x}$ equals k, which is constant. If this ratio held for the coordinates of the origin, it would be $\frac{0}{0} = k$. However, no one value of k would make sense as the value of $\frac{0}{0}$ because the origin is on every line represented by an equation of the form $y = kx$. Thinking about the equation $y = kx$ in terms of number relationships also leads to the conclusion that the value of $\frac{0}{0}$ cannot be determined: If $y = kx$ and $x = 0$, then $y = 0$ and k can be any real number, just as in Focus 1. It is important to note that in the case in which $x = 0$ and $y \neq 0$, such as $\frac{2}{0}$, it is difficult to explain via direct proportion; if $y = kx$, then $x = 0$ and $y \neq 0$ is an impossible circumstance.

A different mathematical context for looking at division involving 0 is the Cartesian product. A nonzero example is this: If 12 outfits can be made using 3 pairs of pants and some number of shirts, how many shirts are there? There must be 4 shirts, as this would give 12 pants–shirt combinations. Similarly, if 0 outfits can be made using 2 pairs of pants and some number of shirts, there must be 0 shirts. If 0 outfits can be made using 0 pairs of pants and some number of shirts, the number of possibilities for the number of shirts is infinite. Lastly, how many shirts are there if there are 2 outfits and 0 pairs of pants? No possible number of shirts can be used to make 2 outfits if there are 0 pairs of pants.

In the context of factor pairs, a division expression with an integral quotient represents an unknown integer factor of the dividend. For $\frac{12}{3}$, 3 and the quotient are a factor pair for 12. In this expression, 12 can be written as the product of 3 and the quotient: $12 = 3 \times 4$. For $\frac{0}{2}$, 2 and the quotient are a factor pair for 0. Therefore, the quotient must be 0, because $0 \times 2 = 0$. For $\frac{0}{0}$, 0 is part of an infinite number of factor pairs for 0 and so the expression is indeterminate. For $\frac{2}{0}$, 0 is not part of any factor pair for 2, thus the expression is undefined.

One side length of a rectangle is the quotient of the area of the rectangle and its other side length. Suppose that rectangles can have side lengths of 0. If a rectangle has area 12 and height 3, what is its width? The width would be 4. If a rectangle has area 0 and length 2, its width is 0, suggesting that 0 divided by 2 is 0. If a rectangle has area 0 and height 0, what is its width? Any width is possible, suggesting that 0 divided by 0 is indeterminate. If a rectangle has area 2 and height 0, what is its width? It is impossible for a rectangle to have area 2 and height 0, suggesting that 2 divided by 0 is undefined.[2]

Mathematical Focus 4

Contextual applications of division or of rates or ratios involving 0 illustrate when division by 0 yields an undefined or indeterminate form and when division of 0 by a nonzero real number yields 0.

If Angela makes 3 free throws in 12 attempts, what is her rate of success? If Angela makes 3 free throws in 12 attempts, her rate is $\frac{1}{4}$. If Angela makes 0 free throws in 2 attempts, her rate is 0. If Angela makes 0 free throws in 0 attempts, her rate could be any of an infinite number of rates. On the other hand, because it is not possible for Angela to make 2 free throws in 0 attempts, it is not possible to determine her rate.

Determining the speed of an object over a given period of time is another rate context. If one travels 12 miles in 3 hours, how fast is one traveling? Traveling 12 miles in 3 hours yields a rate of 4 miles per hour. If one travels 0 miles in 2 hours, one is traveling 0 miles per hour. If one travels 0 miles in 0 hours, how fast is one traveling? An infinite number of speeds are possible. If one travels 1 mile in 0 hours, how fast is one traveling? This situation is impossible because traveling for 0 hours means one is not traveling at all. [Note that there is a sense of infinite speed here, so it might be tempting to define $\frac{1}{0}$ as infinity. However, this leads to further complications, as noted in Focus 5.]

Additionally, the idea of rate is prevalent when calculating unit price, such as when purchasing multiple quantities of an item in a store. If \$12 buys 3 pounds of tomatoes, what is the cost of 1 pound? If \$0 buys 2 pounds of tomatoes, then 1 pound can be bought for \$0. If \$0 buys 0 pounds of tomatoes, there are an infinite number of possible costs for 1 pound. If \$2 buys 0 pounds of tomatoes, it is not possible to determine the number of dollars needed to buy 1 pound.[3]

Mathematical Focus 5

Slopes of lines in two-dimensional Cartesian space map to real projective one-space in such a way that confirms that the value of $\frac{a}{b}$ when b = 0 is undefined if $a \neq 0$ and indeterminate if a = 0.

In the Cartesian plane, consider the set of lines through the origin, and consider each line (without the origin) to be an equivalence class of points in the plane. Except when $x = 0$, the ratio of the coordinates of a point gives the slope of a line—the line that is the equivalence class containing that point. The origin must be excluded because it would be in all equivalence classes, which suggests that $\frac{0}{0}$ would be the slope of any line through the origin [see Focus 3]. Note that the slope of a line through the origin is equal to the y-coordinate of the intersection of that line and the line $x = 1$. The slope then establishes a natural one-to-one correspondence between the equivalence classes (except for the equivalence class that is the vertical line, because it does not intersect the line $x = 1$) and the real numbers. Thus, the real numbers give all possible slopes, except a slope for the vertical line.

When $x = 0$, all points in the equivalence class lie on a vertical line, the y-axis. (Again the origin must be excluded from this equivalence class.) As positively sloped lines approach vertical, their slopes approach ∞, suggesting the slope of the vertical line to be ∞. As negatively sloped lines approach vertical, their slopes approach $-\infty$, suggesting the slope should instead be $-\infty$. However, there is only one vertical line through the origin, so it cannot have two different slopes. To resolve this ambiguity, one might decide that ∞ and $-\infty$ are the same "number" because they should represent the same slope. This set of all possible slopes then consists of all real numbers and one more number, which might be called ∞. Imagine beginning with the extended real line, $\mathbb{R} \cup \{\infty, -\infty\}$, and "gluing together" the points ∞ and $-\infty$ so that they are the same point. This is the real projective one-space, $\mathbb{R} \cup \{\infty\}$.[4]

POSTCOMMENTARY

For situations involving division with 0, there are three types of forms: 0, undefined, and indeterminate. The indeterminate form has particular importance in calculus. Given a function, f, that would be continuous everywhere except that $f(a)$ is indeterminate, a functional value can sometimes be selected to make a related function that is continuous everywhere. For all of its domain values except a, the new function would have the same values as the given function. For example, in the case of the function defined by $f(x) = \frac{\sin x}{x}$ where x is in radians, the function f is continuous for all real numbers except 0, because the functional value at $x = 0$ is the indeterminate form $\frac{0}{0}$. The piecewise-defined function, $f(x) = \begin{cases} \frac{\sin x}{x}, & x \neq 0 \\ 1, & x = 0 \end{cases}$ is continuous for all real numbers. In this case, the fact that the limit of interest was 1 was used: $\lim\limits_{x \to 0} \frac{\sin x}{x} = 1$ where x is in radians. However, in other cases, limits related to $\frac{0}{0}$ do not have to be 1, or even an integer. For example, $\lim\limits_{x \to 0} \frac{2\sin x}{3x} = \frac{2}{3}$. These examples illustrate that, depending on the function, an indeterminate form can sometimes be replaced by a limiting value.

NOTES

1. These types of arguments are ones many students give, and they are important from the point of view of learners being able to attach to this situation meanings with which they already are familiar. Although they are based on real-world ideas (ideas that are informed by starting with a nonzero number of items and breaking those into a non-zero number of groups), the language of dividing collections of real physical objects into groups with fewer objects per group breaks down or becomes meaningless when there are 0 groups. What this indicates is that the metaphor used for attaching everyday meaning to division, namely dividing collections of physical objects into groups with fewer objects per group, is no longer a usable metaphor. The breakdown of this metaphor is that it is not a precise mathematical argument that explains what is problematic about the mathematics used in the missing-factor defini-tion of quotient.

2. See Note 1 for a comment regarding the limitations of these real-world analogies.

3. See Note 1 for a comment regarding the limitations of these real-world analogies.

4. "Gluing together" the ends of the real line creates the same entity as the one-dimensional unit "sphere," namely, the unit circle. That may be easier to picture than real projective one-space and matches the intuition of taking the two supposed "ends" of the real number line, ∞ and $-\infty$, and gathering them up together to form the top "point" (north pole) of the unit circle.

CONNECTION TO STANDARDS

Learning experiences in mathematics for teachers are well positioned when teachers are aware of connections to the standards for which their students are accountable. These standards differ across states and provinces as well as across countries, although there are commonalities across different sets of standards. Although this document cannot possibly address all the existing sets of standards across states, provinces, and countries, an example follows that illustrates how the proposed professional learning might address one particular set of standards. The example addresses the connections of the Common Core State Standards in Mathematics (CCSSM) (National Governors Association Center for Best Practices & Council of Chief State School Officers, 2010) to the proposed professional learning related to the Division Involving Zero Situation. The Common Core State Standards for mathematical content and mathematical practice mirror the attention given in various sets of standards both to what mathematics should be learned and to the mathematical processes in which students should engage. The Appendix of this Guidebook lists the CCSSM Standards for Mathematical Practice, one example of mathematical process standards. Questions (in bold italics in the following chart) that accompany the display of each CCSSM standard can be used in professional learning settings to extend work with specific standards.

Related Common Core Standards
CCSSM Standards for Mathematical Content
Grade 3 Operations and Algebraic Thinking **Represent and solve problems involving multiplication and division.** **3.OA.4.** Determine the unknown whole number in a multiplication or division equation relating three whole numbers. *For example, determine the unknown number that makes the equation true in each of the equations 8 × ? = 48, 5 = ? ÷ 3, 6 × 6 = ?.* ***What happens when 0 is one of the three numbers in the multiplication or division equation?***[3] **Grade 5 Number and Operations—Fractions** **Apply and extend previous understandings of multiplication and division to multiply and divide fractions.** **5.NF.7.** Apply and extend previous understandings of division to divide unit fractions by whole numbers and whole numbers by unit fractions. a. Interpret division of a unit fraction by a non-zero whole number, and compute such quotients. *For example, create a story context for $\frac{1}{3} \div 4$, and use a visual fraction model to show the quotient. Use the relationship between multiplication and division to explain that $\frac{1}{3} \div 4 = \frac{1}{12}$ because $\frac{1}{12} \times 4 = \frac{1}{3}$.*

Why restrict this standard to division by nonzero numbers?

   c.   Solve real world problems involving division of unit fractions by non-zero whole numbers and division of whole numbers by unit fractions, e.g., by using visual fraction models and equations to represent the problem. *For example, how much chocolate will each person get if 3 people share $\frac{1}{2}$ lb of chocolate equally? How many $\frac{1}{3}$-cup servings are in 2 cups of raisins?*

We have $\frac{1}{2}, \frac{1}{3}, \frac{1}{4}$, and so on. Why not $\frac{1}{0}$?

Grade 7 The Number System

Apply and extend previous understandings of operations with fractions to add, subtract, multiply, and divide rational numbers.

7.NS.2.   Apply and extend previous understandings of multiplication and division and of fractions to multiply and divide rational numbers.

   b.   Understand that integers can be divided, provided that the divisor is not 0, and every quotient of integers (with non-zero divisor) is a rational number. If p and q are integers, then $-\frac{p}{q} = \frac{-p}{q} = \frac{p}{-q}$. Interpret quotients of rational numbers by describing real world contexts.

Why restrict this standard to nonzero divisors?

High School—Number and Quantity

Use properties of rational and irrational numbers.

N.RN.3.   Explain why the sum or product of two rational numbers is rational; that the sum of a rational number and an irrational number is irrational; and that the product of a nonzero rational number and an irrational number is irrational.

Is the product of a whole number and an irrational number always an irrational number? Explain.
Is the sum of two irrational numbers always an irrational number? Explain.

High School—Algebra

Rewrite rational expressions.

A-APR.7. (+)   Understand that rational expressions form a system analogous to the rational numbers, closed under addition, subtraction, multiplication, and division by a nonzero rational expression; add, subtract, multiply, and divide rational expressions.

How does understanding division by 0 with rational numbers help us work with rational expressions?
The following is considered an identity: $(x - 1) \cdot (x + 1) = x^2 - 1$. Can the following be considered an identity: $\frac{x^2 - 1}{x - 1} = x + 1$?

CCSSM Standards for Mathematical Practice[4]

SMP2.   Reason abstractly and quantitatively.
SMP3.   Construct viable arguments and critique the reasoning of others.
SMP6.   Attend to precision.

SUGGESTIONS FOR USING THIS SITUATION

Facilitators may want to peruse what follows for an idea about how to use the Division Involving Zero Situation in their professional learning settings. The chart provides an Outline of Participant activities and a summary of the Tools and projected Time required for implementation of those activities. Following the outline are Facilitator Notes that describe each of the suggested activities in greater detail.

Tools	Time
• A range of (at least 5 different) calculators and/or other computing technologies (e.g., spreadsheet, smartphone applet) that give different results for division involving 0 • Graphing utility (e.g., graphing calculator, dynamic geometry system, online computer algebra system) • Poster paper, markers • Copies of the Prompt (separate from the Foci) • Copies of the Foci	2–3 hours, can be done in a single session or across multiple sessions
Outline of Participant Activities (Details for these activities follow in the Facilitator Notes section.)	

Launch	The Launch for the proposed professional learning session(s) problematizes division by 0. Participant activities in the Launch bring to the fore issues that arise as teachers and students encounter various indicated divisions involving the number 0.
Activity 1.	Participants read the Prompt, think about their own answers for each problem in the Prompt ($2 \div 0 = ?$; $0 \div 0 = ?$, $0 \div 2 = ?$), and then discuss their answers to each problem with group members or with a partner.
Activity 2.	Using different computing technologies, participants explore answers for each problem from the Situation's Prompt.
Activity 3.	Participants reevaluate original responses in light of answers obtained from technologies and consider mathematical ideas that could be used to support the answers that participants now consider to be correct.
Activity 4.	Participants analyze and discuss the Foci, then structure an argument to present to the entire group.
Reflect and assess learning	Participants reflect on and discuss how division involving 0 might come up in their classrooms–how they might respond and how they might foster student consideration of division involving 0. Suggested activities incorporate ways for participants to assess their understanding.

FACILITATOR NOTES

About the mathematics

Distinguish between *undefined* and *indeterminate*. Often in mathematics an assumption is that when a value is excluded (e.g., $n \neq 0$) the value of the expression using the excluded value is undefined. However, values are sometimes excluded because they lead to an indeterminate expression. The expression $\log_b x$ is an example in which undefined and indeterminate both occur. When $b = 0$, the expression $\log_b x$ is undefined for all values of x other than 0, because 0 raised to any power will never be a value other than 0. When $b = 0$ and $x = 0$, the expression $\log_b x$ is defined but indeterminate, because 0 raised to any positive power equals 0, and so $\log_b 0$ does not have a unique value.

Launch

Time

15–20 minutes

In the Launch, establish the importance of this idea and the need for teachers to understand the underlying mathematics so that they can help students understand division involving 0 and not just give students "the rules." There may be a subset of participants who do not think this is important to clarify, that they are currently addressing it sufficiently, or that giving the rule is sufficient. To address this, ask participants why understanding this is important and where this idea occurs in the mathematics curriculum. Possible occurrences they may identify include:

- Basic division facts: What is $0 \div 3$? $3 \div 0$? $0 \div 0$?

- The definition of rational number is $\dfrac{a}{b}$ where a and b are integers, $b \neq 0$.

- The slope of a vertical line is undefined.

- The product of a nonzero rational number and an irrational number is an irrational number.

- Throughout mathematics, statements often have exclusions involving 0.

The Related Common Core Standards chart in the Connection to Standards section identifies where division involving 0 occurs in the Common Core State Standards for Mathematics.

Be sure that participants understand a variety of mathematical notation for division, for example, that when $\dfrac{2}{0}$ indicates division, it should be read as *2 divided by 0*.

Activity 1. Participants read the Prompt, think about their own answers for each problem in the prompt (2 ÷ 0 = ? , 0 ÷ 0 = ? , 0 ÷ 2 = ?), then discuss their answers to each problem with their group members or with a partner.

Ask participants whether these explanations in the Prompt have ever come up, or might come up, in their classroom and how they have addressed them or would address them with students.

Time

7–10 minutes

Anticipated participant responses

In addition to the responses given in the Prompt, participants may think there are multiple answers, for example, $0 ÷ 0$ is sometimes 0, sometimes 1; or that anything divided by 0 is 0 (Ball, 1990).

Facilitating the activity

Let participants discuss in their groups; however, do not have groups report out. Chances are at least some of the participants' responses will be incomplete or incorrect. The purpose is for participants to reflect on their current knowledge of division involving 0 and make different ideas public, including incomplete conceptions, rather than to reach conclusions. This is an opportunity for you to assess the mathematics background of different groups and determine which Focus might be most useful for each group to analyze subsequently in Activity 4.

Any of the following ideas might arise from a group:

- A rule; that's the way it is
- Concrete situations (Foci 2 & 3)
- Equations (Focus 1)
- Division as partitioning (Focus 2)
- Rational numbers as division (Focus 3)
- Ratio and rate (Focus 4)
- Slope (Focus 5)

Activity 2. Explore answers for each problem obtained from different computing technologies.

Time

10 minutes

Anticipated responses from various computing technologies

See Table 2.1 at the end of this chapter for responses from selected computing technologies.

Facilitating the activity

The purpose is for participants to recognize that different technologies provide different results for these various computations, so that technology will not resolve the issue for them. Consequently, this is a problem worth analyzing in depth. This also illustrates that results obtained from computing technologies are not always correct. Technology is not the focus of the activity, so do not spend time trying to explain how the technology arrives at different results.

Do not spend much time on this activity. In fact, this activity may be unnecessary if participants use technology to determine their answers in Activity 1.

Assign each pair or group of participants to get the answers to the three problems from one or two computing technologies, depending on the size of the group and number of technologies to investigate. We recommend looking at the results from at least five different technologies. Be sure to try out the technologies yourself; even though we found the different results in Table 2.1, technology programs change, so current results may differ from those listed. Participants may want to try their personal digital tools, which is fine. The intent is for the collective set of technologies to give conflicting results. Create a list of the different results for each problem, similar to those in Table 2.1. Participants may want to discuss the ways in which technology can help or hinder investigation of division involving 0.

Another approach is to have a technology scavenger hunt; for example, challenge each pair or group to find as many different answers to these problems as possible. Be prepared with sources that provide different answers (such as those listed in the table) in case participants do not find a range of answers.

Activity 3. Reevaluate original responses in light of answers obtained from different technologies and consider mathematical ideas that could be used to support the answers they now think are correct.

Time

5 minutes

Anticipated participant responses

Participants might interpret division involving 0 in terms of:

- A rule; that's the way it is

- Concrete situations (Foci 2 & 3)

- Equations (Focus 1)

- Division as partitioning (Focus 2)

- Rational numbers as quotients (Focus 3)

- Ratio and rate (Focus 4)

- Slope (Focus 5)

Facilitating the activity

The purpose is to learn the mathematics about which each group is thinking and with which participants are familiar. This may aid in assigning a Focus to each group, if that is the approach to be taken.

Ask each group to discuss the mathematics they would use to support their solutions. Collect ideas from each group on one poster per group.

Activity 4. Analyze and discuss the Foci, then structure an argument to present to the entire group.

Time

1.5–2 hours

Facilitating the activity

This activity may be structured in a number of ways, depending on the mathematical understanding of the participants, their instructional level (elementary school, middle school, or high school), and the amount of time available. The amount of emphasis on particular Foci will also depend on participants' understanding and background.

Use this prompt for each group:

- Analyze your assigned Focus or Foci.

- Create a 3- to 7-minute explanation using this mathematical Focus (or Foci) to explain the three cases involving division involving zero—$0 \div 2$, $2 \div 0$, and $0 \div 0$—to the whole group or as if to the pre-service middle school teachers in the Situation Prompt.

Option 1: All groups analyze all Foci or subsets of Foci

Assign all Foci or the same subset of Foci to each group. Groups should be prepared to present any of the assigned Foci. Consider asking participants to compare/contrast the Foci (what's similar, what's different), and/or rank the Foci in terms of which they find most compelling or transparent for students whom they teach. (Note: Asking participants to rank items often produces richer discussion than simply asking them to discuss or analyze each one.) Also consider asking how each Focus advances the participants' own understanding of division involving 0.

Option 2: Each group analyzes only one Focus or a subset of Foci

Assign a different Focus or different sets of Foci to each group. In doing so, take into consideration what you know about their mathematics background and analysis of division involving 0 in Activities 1, 2, and 3. Each group should prepare a presentation about its Focus. While listening to presentations of other groups about their Foci, participants should be asked to compare and contrast their Focus with that of others, and also consider how each Focus advances their own understanding of division involving 0.

Key points about the Foci

- Note the difference between $n \div 0$ ($n \neq 0$) and $0 \div 0$. The indicated quotient, $n \div 0$, is *undefined*—there is no solution to the corresponding multiplication equation; $0 \div 0$ is *indeterminate*—there are infinitely many solutions to the corresponding multiplication equation. Even though textbooks typically refer to division by 0 as being "undefined," use of this term to describe the case $0 \div 0$ is mathematically imprecise.

- Focus 1 addresses division involving 0 through corresponding multiplication equations.

- Focus 2 addresses division involving 0 through the two different interpretations of whole number division: finding the number of objects in each equal-sized group (partitive division) or finding the number of equal-sized groups (quotitive division). The point is not the vocabulary associated with the two interpretations of division, but to use both interpretations to analyze division involving 0.

- Focus 3 addresses division involving 0 through five interpretations of $\frac{a}{b}$ as a rational number: slope, direct proportion, Cartesian product (combinations), factor pair (if $ab = c$, then a and b form a factor pair for c), and area of rectangle. Each interpretation is applied to address division involving 0. Depending on the time available and the number of participants, consider assigning different groups to analyze selected interpretations in this Focus.

- Focus 4 uses rates as a context for analyzing division involving 0. Three specific rates are used: success rate, distance per time (e.g., mph), and unit price. Although participants may have different definitions of *rate* and *ratio*,[5] the distinction between these terms is not the point of this Focus. Keep participants' attention on the use of rates and ratios to analyze division involving 0. You might ask participants to create their own rate example, in addition to analyzing the three provided.

- Focus 5 maps the slopes of lines in two-dimensional Cartesian space to real projective one-space to analyze division involving 0. This Focus requires that participants know what a real projective space is, as well as understand equivalence classes. Consequently, this Focus may not be accessible to all teachers. On the other hand, this Focus highlights mathematics that secondary teachers may want to explore further.

- The Postcommentary provides a discussion that connects this topic to limits in calculus.

Key points for discussion

- After presentations, ask participants to compare and contrast the ways that division involving 0 is addressed in the different Foci. For example, ask how the idea of indeterminate form was brought out in the contexts presented in the different Foci.

  Likely connections include those between Focus 1 and Focus 3 (viewing division through a multiplicative lens).

- Engage participants in a reflective discussion about their learning from this activity:

  ○ Which Foci had you previously considered?

  ○ Which were new to you?

  ○ Which ones were most compelling?

  ○ What new insights did you gain, if any, from looking at division involving 0 from these different Foci?

  ○ What confusion remains about division involving 0?

Reflect and assess learning

The purpose for this reflection is for participants to connect the session's activities to their own classroom practice and to assess what participants learned from the session. Specific activities depend on the setting for the professional learning session, that is, whether it is a stand-alone session or part of an ongoing series of sessions.

Time

45–60 minutes

Suggested assessment activities

- Ask participants to determine when division involving 0 might come up in their curriculum, then design and/or adapt a lesson/task/explanation to ensure that their students understand division involving 0 in this context. Participants would work in grade-level groups and make a poster of their lesson/task/explanation. If this session is part of an ongoing series of professional development sessions, participants could be asked to try their task/lesson/explanation in their classroom, then bring classroom artifacts to the next session, as a lesson study or modified lesson study.

One way to debrief and evaluate participants' lessons is to do a modified gallery walk, in which, as participants read/review each lesson, they write questions about the lesson on "sticky" notes. As the facilitator, you might want to provide "expert commentary" on the lessons in addition to participants' comments. Participants could be asked to revise their lesson based on the feedback they received.

Another possibility is to structure a participant peer-review process in which the most compelling arguments are identified. One option is a process of elimination: (a) lessons are paired, and subsets of participants select the stronger of each pair of lessons; (b) the selected lessons are then compared, and so on, until one lesson (or one per grade band) emerges as the strongest explanation. Another option is to ask each participant to vote on the three explanations that they find the most compelling.

- Ask participants to revisit the Situation's Prompt and determine how they would respond to students who gave each of the answers to $0 \div 0$, for example, "How would you respond to a student who said $0 \div 0$ is equal to 1 because a number divided by itself is 1?"

  This activity could be an individual writing prompt, or different prompts could be given to different groups. The previously described options for debriefing could be used here.

- Also consider asking participants to reflect on which of the Standards for Mathematical Practice in which they were engaged during the session.

Suggested reflection questions

1. Has our work with this Situation caused you to consider or reconsider any aspects of your own thinking and/or practice about division involving 0? Explain.

2. Has our work with this Situation caused you to reconsider any aspects of your students' mathematical learning about division involving 0? Explain.

3. What additional questions has our work with this Situation raised for you?

RESOURCES

Reys and Grouws (1975) might be used to expand on how to access children's thinking about division involving 0. It offers insights into children's responses to four division expressions: $12 \div 3$, $0 \div 4$, $8 \div 0$, and $0 \div 0$.

Knifong and Burton (1980) offers five intuitive approaches—repeated subtraction, equivalent sets, arrays, the number line, and a mathematical balance—to division expressions in which 0 is the dividend, the divisor, or both the divisor and the dividend. Note that these authors imprecisely refer to the infinitude of possible quotients for 0 ÷ 0 as indicating that 0 ÷ 0 is *undefined* rather than *indeterminate*.

Although the Grouws and Reys (1975) article likely would not be used with participants in a professional learning setting, it offers background information on an experimental study with 755 students in Grades 4, 6, and 8 who were given instruction based on the inverse relationship between multiplication and division and tested on division items, two thirds of which involved 0 as either the divisor or the dividend, but not both.

NOTES

1. The Division Involving Zero Situation is one of the Situations presented in *Mathematical Understanding for Secondary Teaching: A Framework and Classroom-Based Situations* (Heid & Wilson, 2015).

2. This Situation appears on pp. 95–101 of Heid and Wilson (2015). It is reprinted with permission.

3. For each standard, we have suggested questions (those in italicized, boldface type) that might serve as follow-up questions to address the mathematics of the standard.

4. See the Appendix for the list of the Standards for Mathematical Practice in the Common Core State Standards for Mathematics (National Governors Association Center for Best Practices and Council of Chief State School Officers, 2010).

5. Ratios can be used to make part-to-whole, part-to-part and whole-to-whole comparisons. Rates are typically defined as comparisons of quantities with different units (e.g., miles per hour, number of completions per number of attempts). For more information, see Lobato and Ellis (2010).

REFERENCES

Ball, D. L. (1990). Prospective elementary and secondary teachers' understanding of division. *Journal for Research in Mathematics Education, 21*, 132–144.

Grouws, D. A., & Reys, R. E. (1975). Division involving zero: An experimental study and its implications. *The Arithmetic Teacher, 22*, 74–80.

Heid, M. K., & Wilson, P. W. (with G. W. Blume). (Eds.). (2015). *Mathematical understanding for secondary teaching: A framework and classroom-based situations*. Charlotte, NC: Information Age.

Knifong, J. D., & Burton, G. M. (1980). Intuitive definitions for division with zero. *Mathematics Teacher, 73*, 179–186.

Lobato, J., & Ellis, A. B. (2010). *Developing essential understanding of ratios, proportions, and proportional reasoning for teaching mathematics in grades 6–8*. Essential understanding series. (R. Charles, Vol. Ed.; R. M. Zbiek, Series Ed.). Reston, VA: National Council of Teachers of Mathematics.

National Governors Association Center for Best Practices & Council of Chief State School Officers. (2010). *Common core state standards for mathematics*. Washington, DC: Authors.

Reys, R. E., & Grouws, D. A. (1975). Division involving zero: Some revealing thoughts from interviewing children. *School Science and Mathematics, 75*, 593–605.

TABLE 2.1. Values for 0 ÷ 2, 0 ÷ 0, and 2 ÷ 0 Displayed by Various Technological Tools (Retrieved as of May 19, 2017)

Tool	0 ÷ 2	0 ÷ 0	2 ÷ 0
Calculators			
Casio fx-CG10	0	Ma ERROR	Ma ERROR
TI-89	0	undefined	undefined
TI-Explorer Plus	0	Error Ari	Error Ari
Calculator Apps			
iPad app (HD calculator)	0	0	0
iPad app (pocketCAS)	0	undef	∞
Mac computer (calculator application)	0	Not a number	Not a number
Phones			
iPhone	0	error	error
Calculator on Android phone	0	Invalid operation	Invalid operation
Software (other than calculator apps)			
Excel spreadsheet	0	#DIV/0!	#DIV/0!
Geometer's Sketchpad	0	undefined	∞
Demos Scientific Calculator	0	undefined	undefined
Websites			
http://mathworld.wolfram.com/DivisionbyZero.html	(not addressed by this site)	The uniqueness of division breaks down when dividing by zero, ... division by zero is undefined for real numbers ...	The uniqueness of division breaks down when dividing by zero, ... division by zero is undefined for real numbers ...
Wolfram Alpha	0	(undefined), followed by reference to "other indeterminate forms"	$\tilde{\infty}$ Complex infinity "Complex infinity is an infinite number in the complex plane whose complex argument is unknown or undefined. Complex infinity may be returned by the Wolfram Language, where it is represented symbolically by ComplexInfinity. The Wolfram Functions Site uses the notation "infinity overscored by ~" to represent complex infinity.
http://en.wikipedia.org/wiki/Division_%28mathematics%29 see *Division (mathematics)* entry, *By zero* section	(not addressed by this site)	Division of any number by zero (where the divisor is zero) is undefined.	Division of any number by zero (where the divisor is zero) is undefined.
http://en.wikipedia.org/wiki/Division_by_zero	(not addressed by this site)	... division by zero is undefined. Since any number multiplied by zero is zero, the expression 0/0 has no defined value and is called an indeterminate form.	... division by zero is undefined.

Tool	0 ÷ 2	0 ÷ 0	2 ÷ 0
Ask Dr. Math, Dr. Robert's answer to "Why is 0÷0 'indeterminate' and 1/0 'undefined'?"		Division by zero is an operation for which you cannot find an answer, so it is disallowed.	
HandwritingForKids.com/ handwrite/math/division/ mathfacts.htm (See table of values)	0	0	(not addressed by this site)
WEB 2.0 CALC (https:// web2.0calc.com/)	0	Error: DivByZero	Error: DivByZero
Scientific calculator at Math. com	0	error	error
www.calculator.com (beta version)	0	NAN (presumably an abbreviation for *not a number*)	∞

CHAPTER 3

FACILITATOR'S GUIDE
FOR
PRODUCT OF TWO NEGATIVE NUMBERS

Situation 2 From the MACMTL-CPTM Situations Project[1]

Glendon Blume and Connie Schrock

The Situation selected for this Guide is one that resonates with the experience of many teachers. Every mathematics teacher at one time or another is faced with addressing why it is reasonable for the product of two negative numbers to be a positive number. Students may ask why, textbooks may state such a "rule," or reasoning about a product may require analysis of signs. The following pages offer ideas for exploring in a professional learning or teacher preparation context a Situation about the product of two negative numbers. Although the Situation was developed from a teaching dilemma in a secondary classroom, its issue is faced by elementary teachers as well.

Facilitator's Guidebook for Use of Mathematics Situations in Professional Learning,
pages 29–58.

OVERVIEW

Facilitators can scan the following overview to quickly get a sense of the mathematics involved in the proposed professional learning setting. The sections deal with the mathematics of the Product of Two Negative Numbers Situation, why these mathematical ideas might be important for participants, the learning goals for participants, and mathematical ideas central to the proposed professional learning sessions.

Situation	Relevance
The question "Why is it that when you multiply two negative numbers together, you get a positive number answer?" is a common query in middle school and high school classrooms. This Situation offers several ways to think about the product of two negative numbers, including ones based on patterns in tables of values, transformations of vectors, geometric analogies, and properties of the real number system.	• Preservice and in-service teachers at all levels are faced with the question of why the product of two negative numbers is positive. They are likely to "know" that the product of two negative numbers is positive but may have difficulty explaining why that is the case. • Some explanations for why the product of two negative numbers is positive are analogies that are intuitively compelling and thus are used to support this result, although they are not proofs. Others—based on assumptions about and properties of the real number system—constitute proofs of that result. Teachers will find it useful to be able to distinguish analogical explanations from proofs.
Goals	**Key Mathematical Topics**
• Participants will develop understanding of a variety of ways to convince students that the product of two negative numbers is positive. • Participants will develop an understanding of the difference between an intuitive argument about the reasonableness of a result and a proof of that result.	• The use and limitations of a repeated-addition model for multiplication applied to products of negative numbers • Real-world settings that have results (two negatives produce a positive) analogous to multiplication of negative numbers • Recognition of the mathematical limitations of some of these settings in that they do not embody the operation of multiplication nor do they logically support the conclusion that the product of two negative numbers is a positive number • Multiplication of a one-dimensional unit vector by the scalar -1 as an analogy to multiplying a real number by a negative number • Use of the distributive property of multiplication over addition to re-express products to develop intuitive arguments about the sign of a product of negative numbers • Use of regularity in the values of real-valued functions to support an intuitive argument about the sign of a product of negative numbers • The difference between an intuitive argument and a proof • The use of assumptions about the real number system and its properties as the basis for a proof that the product of two negative numbers is positive

The Situation under consideration in this chapter follows (highlighted with a gray background).[2]

PRODUCT OF TWO NEGATIVE NUMBERS

Situation 2 From the MACMTL–CPTM
Situations Project

Ryan Fox, Sarah Donaldson, M. Kathleen Heid,
Glendon Blume, and James Wilson

PROMPT

A question commonly asked by students in middle school and secondary mathematics classes is "Why is it that when you multiply two negative numbers together, you get a positive number answer?"

COMMENTARY

Students are able to visualize the addition and subtraction of integers, but multiplication of integers, particularly signed numbers, seems to be more abstract. Representing multiplication of quantities less than 0 is difficult. The Foci make this abstract concept more concrete by providing multiple ways to think about the multiplication of negative numbers, some of which only suggest that the product of two negative numbers should be positive and some of which establish definitively that the product of two negative numbers is positive via a general proof. Focus 1 applies a repeated addition model to multiplication of negative numbers, but that model is shown to have limitations. Real-world applications often are used to suggest that the product when multiplying two negative numbers should be positive; Focus 2 offers one such application involving an employee's pay. Focus 3 employs a visual approach using scalar properties of vectors to suggest that the product should be positive. In Focus 4, the distributive property and other properties of the real numbers are used to show, for specific cases and in general, that the product of two negative numbers should be a positive number. Focus 5 develops an intuitive, pattern-finding approach to suggest that the product should be positive. A geometric argument based on similar triangles appears in Focus 6. Focus 7 offers an analysis based on some concepts from abstract algebra.

MATHEMATICAL FOCI

Mathematical Focus 1

Repeated addition suggests that the product of a negative integer and any negative number is a positive number.

Repeated addition is one way in which whole number multiplication typically is introduced. The product 3×5 can be thought of as adding 5 three times (starting from 0): $0 + 5 + 5 + 5$, or 15. Similarly, multiplication involving a negative integer, for example, 3×-5, can be thought of as repeated addition: Starting from 0, there are three addends, each of which is -5: $0 + (-5) + (-5) + (-5)$. Addition of negative integers leads to the result $3 \times -5 = -15$. Due to the commutative property of multiplication, the product -5×3 also equals -15. However, the repeated addition model for the product -3×-5 is not straightforward because it is difficult to interpret the product as -3 addends, each of which is -5. In Table 8.1, in moving from one row to the row below it (decreasing the first factor in the product by 1), an addend of -5 is being taken away from (subtracted from) the sum. The pattern suggests that the entry after 0 would be computed by subtracting an addend of -5 from 0, which is the same as adding the opposite of -5, meaning that 5 is the result, and suggesting that the product of two negative integers is a positive integer.

If one factor is negative but not an integer, the repeated addition model can be applied. For example, if the product in Table 8.1 had been -3×-5.1 (or even -5.1×-3, which is equal to -3×-5.1 because real number multiplication is commutative), each addend would have been -5.1, but the number of addends and the signs of the products would remain the same.

The repeated addition model for multiplication of negative numbers can suggest that the product of a negative integer and another negative number is positive; however, the model is not helpful when both factors are negative and neither is an

TABLE 8.1   The Sum That Results When -5 Is Used a Decreasing Number of Times as an Addend

Product	Number of addends	Sum
3×-5	3 addends, each of which is -5	$(-5) + (-5) + (-5)$, or -15
2×-5	2 addends, each of which is -5	$[(-5) + (-5) + (-5)] - (-5)$, or -10
1×-5	1 addend, which is -5	$[(-5) + (-5)] - (-5)$, or -5
0×-5	0 addends, each of which is -5	$[(-5)] - (-5)$, or 0
-1×-5	One fewer addend, or -1 addends, each of which is -5	$[0] - (-5)$, or $0 + (+5)$, which is 5
-2×-5	-2 addends, each of which is -5	$[0 - (-5)] - (-5)$, or $0 + (+5) + (+5)$, which is 10
-3×-5	-3 addends, each of which is -5	$[0 - (-5) - (-5)] - (-5)$, or $0 + (+5) + (+5) + (+5)$, which is 15

integer. For example, if one multiplies -3.7 × -5.1, one would need to interpret what is meant by -3.7 addends, each of which is -5.1. Also, if both factors were negative irrational numbers, the repeated addition model would require one to interpret what was meant by an irrational number of addends. Although the repeated addition model for multiplication by a negative number is illuminating for negative integer multiplication and suggests—but does not prove—that the product of any negative integer and any other negative number is positive, it should not be interpreted to imply that the product of any two negative numbers is positive.

Mathematical Focus 2

Real-world instances that involve adding or subtracting positive or negative amounts can be used to suggest that the product of two negative numbers is a positive number.

Debts, debits, or deductions, as well as savings, credits, or deposits, can be used to illustrate that the product of two negative numbers is a positive number. Removal or deduction of an amount from an employee's paycheck is negative, whereas addition of an amount to an employee's paycheck is positive. Suppose that an employer deducts $120 per month from an employee's paycheck for health insurance. After 6 months, the deduction (debit) is 6 × (-$120), or -$720. Multiplication models the total amount.

If the employer were to offer the employee a special benefit of removing those deductions (debits), that removal of a negative amount would be a negative action on a negative value. Just as implementing the six deductions can be modeled by 6 × (-$120), removal of six -$120 deductions can be modeled by -6 × (-$120). Removal of those deductions would result in a gain in pay, over 6 months, of $720 for the employee, suggesting that (-6) × (-$120) = $720. Thus, a negative number of negative transactions—a negative number times a negative number—results in a positive number of dollars in pay.

Mathematical Focus 3

Products of negative numbers can be represented as the composition of two reflections.

Vectors on a number line can offer insight into the product of two negative numbers. When a vector on the number line is multiplied by the scalar -1, the vector is reflected about the origin. Consider the vector j in Figure 8.1. Multiplication of j by the scalar -1 yields the vector j', or $-j$. A second reflection, resulting from multiplication of this new vector, j' or $-j$, by the scalar -1, would yield its reflection about the origin, the vector $-j'$ or $-(-j)$ as illustrated in Figure 8.2.

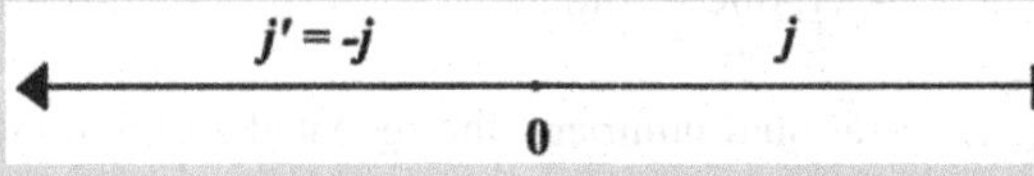

FIGURE 8.1.   The vectors j and –j.

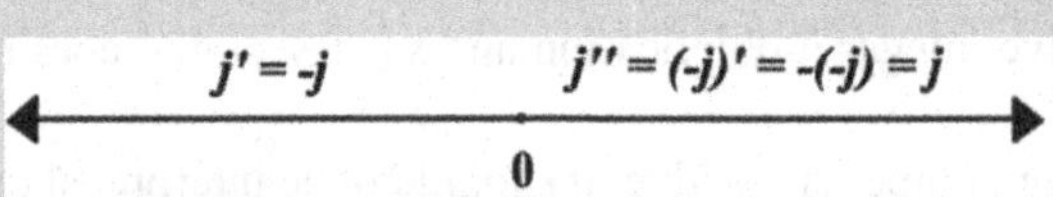

FIGURE 8.2.   The vectors –j and –(–j).

The product of the two reflections is the original vector, j, that is

$$(-1)[(-1)\,j] = 1\,j.$$

But by the associative property of scalar multiplication,

$$(-1)[(-1)\,j] = [(-1)\,(-1)]\,j,$$

suggesting that for real numbers, $(-1)\,(-1) = 1$. Extending this to any two negative scalars suggests that the product of those two scalars, or the product of two negative numbers, is positive.[1]

Mathematical Focus 4

The distributive property of multiplication over addition can be used to illustrate and justify that the product of two negative numbers is a positive number.

The distributive property of multiplication over addition, $a(b + c) = ab + ac$, enables one to write the product of a number and a sum as the sum of two products, which facilitates writing the product $(a + b)\,(c + d)$ as $ac + bc + ad + bd$. Starting from the product $7 \times 3 = 21$, one can write

$$(11 - 4)(5 - 2) = 21.$$

Writing subtraction as addition of the opposite yields

$$[11 + (-4)][5 + (-2)] = 21.$$

Applying the distributive property successively yields

$$[11 \times 5] + [(-4) \times 5] + [11 \times (-2)] + [(-4) \times (-2)] = 21.$$

Given that one has previously established that the product of a negative number and a positive number is a negative number,

$$[55] + [-20] + [-22] + [(-4) \times (-2)] = 21, \text{ or}$$

$$13 + [(-4) \times (-2)] = 21$$

Solving for the term $[(-4) \times (-2)]$ produces

$$[(-4) \times (-2)] = 8,$$

providing an example of the product of two negative numbers being a positive number. An infinite number of such examples is possible, each involving use of the distributive property after rewriting each factor as a difference. In each case, one can use factors of known products (in this case, 7 and 3 as factors of 21) to find the value of the product of two negative numbers.

Other illustrations of the use of properties of the real numbers are possible.[2] For example, the distributive property allows one to determine the value of the expression $2 \times 3 + (-2) \times 3 + (-2) \times (-3)$ in two different ways, and doing so can help to establish that $(-2) \times (-3) = 6$.

$$
\begin{aligned}
2 \times 3 + (-2) \times 3 + (-2) \times (-3) \quad &= 2 \times 3 + (-2) \times [3 + (-3)] \quad \text{factoring out -2),} \\
&= 2 \times 3 + (-2) \times [0] \quad \text{because } a + (-a) = 0, \\
&= 2 \times 3 + 0 \quad \text{because } a \times 0 = 0, \\
&= 2 \times 3 \quad \text{because } a + 0 = a.
\end{aligned}
$$

Taking the same expression and factoring out (3) from the first two terms:

$$
\begin{aligned}
2 \times 3 + (-2) \times 3 + (-2) \times (-3) \quad &= [2 + (-2)] \times 3 + (-2) \times (-3) \quad \text{factoring out 3} \\
&= [0] \times 3 + (-2) \times (-3) \quad \text{because } a + (-a) = 0, \\
&= 0 + (-2) \times (-3) \quad \text{because } 0 \times a = 0, \\
&= (-2) \times (-3) \quad \text{because } 0 + a = 0.
\end{aligned}
$$

So, in the one case the expression equals 2×3, and in the other it equals $(-2) \times (-3)$. The symmetric and transitive properties of equality then guarantee that $(-2) \times (-3)$ is the same number as 2×3, illustrating that the product of these two negative numbers, -2 and -3, is the positive number, 6. However, this argument can be generalized to establish that the product of any two negative numbers is a positive number.

For two positive real numbers, a and b, consider rewriting the expression $ab + (-a)(b) + (-a)(-b)$ in two different ways.

$$ab + (-a)(b) + (-a)(-b) \quad = ab + (-a)[b + (-b)] \quad \text{factoring out } -a,$$
$$= ab + (-a)(0) \quad \text{because } x + (-x) = 0,$$
$$= ab + 0 \quad \text{because } -x \times 0 = 0,$$
$$= ab \quad \text{because } x + 0 = x.$$

And,

$$ab + (-a)(b) + (-a)(-b) \quad = [a + (-a)](b) + (-a)(-b) \text{ factoring out } b,$$
$$= (0)b + (-a)(-b) \quad \text{because } x + (-x) = 0,$$
$$= 0 + (-a)(-b) \quad \text{because } 0 \times x = 0,$$
$$= (-a)(-b) \quad \text{because } 0 + x = x.$$

By the symmetric and transitive properties of equality, $(-a)(-b) = ab$. If a and b are positive real numbers as assumed initially, $(-a)(-b)$ is the product of two negative numbers because the additive inverse of each of those positive numbers is a negative number. Also, ab is positive because it is the product of two positive numbers. Therefore, the product of any two negative numbers is a positive number.

Mathematical Focus 5

Investigating patterns in real-valued functions yields insight into the product of two negative numbers.

A table of values for a function of the form $f(x) = ax$, where $x \geq 0$ (see Table 8.2 for an example using $a = -5$) offers a way to develop a pattern involving the product of two numbers. The pattern established in Table 8.2 is that as the value of x decreases by one, the value of negative 5 times x increases by 5. As the value of x continues to decrease from positive numbers, through 0, to negative numbers (see Table 8.3), the value of $f(x)$ increases first from negative numbers to 0. The increase of 5 in $f(x)$ with each decrease of 1 in x can be established by examining the rate of change of $f(x)$. Because $f(x)$ is a linear function, its rate of change (slope of its graph) is a constant, -5, or $\frac{-5}{1}$. This means that the value of $f(x)$ decreases by 5 for each increase of 1 in x. If this pattern in the function values continues to hold, then $f(x)$ will be positive when x is

TABLE 8.2   Values of $f(x) = -5x$ for Positive and Zero Values of x

x	$f(x) = -5x$	$f(x)$
4	$f(x) = -5 \times 4$	-20
3	$f(x) = -5 \times 3$	-15
2	$f(x) = -5 \times 2$	-10
1	$f(x) = -5 \times 1$	-5
0	$f(x) = -5 \times 0$	0

TABLE 8.3   Values of $f(x) = -5x$ for Positive, Zero, and Negative Values of x

x	$f(x) = -5x$	$f(x)$
1	$f(x) = -5 \times 1$	-5
0	$f(x) = -5 \times 0$	0
-1	$f(x) = -5 \times -1$	5
-2	$f(x) = -5 \times -2$	10

negative. For $-5 \times -2 = 10$, the product of two negative numbers is a positive number. The pattern suggests that this will be true for any product of -5 and a negative number.[3]

Mathematical Focus 6

A geometric model based on similar triangles can suggest that the product of two negative numbers is a positive number.

Suppose that two lines intersect at a common point labeled as 0 for each line. From the definition of a negative number as the additive inverse or opposite of a positive number, label positive and negative directions on each line. Using the multiplicative identity, 1, indicate a unit in each direction on each line (see Figure 8.3). Any point on either line has an orientation, positive or negative, determined by its location.

Positive number times a positive number

Begin by considering a positive value, a, on the horizontal line and a positive value, b, on the slanted line. Next construct a line segment from 1 on the horizontal line to b on the slanted line. Construct a parallel line segment from a on the horizontal line to its intersection with the slanted line. From the geometry of similar triangles, the location of the intersection is the product ab (see Figure 8.4). This is also the classic construction of the fourth proportion, $1:b = a:x$, which leads to $x = ab$.

Because the triangles are similar, the location of the intersection represents the product of a and b. This case suggests that the product of two positive numbers is positive.

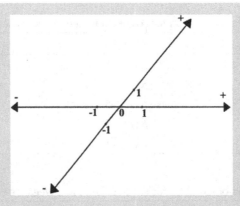

FIGURE 8.3. Two intersecting lines with unit distances marked in both positive and negative directions.

Negative number times a positive number

Consider the preceding construction but beginning with *a* located in the negative direction and *b* located in the positive direction (see Figure 8.5). The construction proceeds in the same way, with *a* on the horizontal, *b* on the slant line, and construction of the segments from the 1 location to the *b* location and then a parallel segment from the *a* location to the intersection point. Again, the intersection point will be the product *ab*, but now this product is located in the negative direction. This suggests that the product of a negative number and a positive number is a negative number. Again this is a direct consequence of the similar triangles.

Positive number times a negative number

Consider the previous construction but beginning with *a* located in the positive direction and *b* located in the negative direction (see Figure 8.6). The construction is done in the same way, with *a* on the horizontal, *b* on the slanted line, and construction of the segments from the 1 location to the *b* location and then a parallel segment from the *a* location to the intersec-

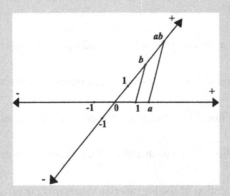

FIGURE 8.4. The product *ab* is positive when both *a* and *b* are positive.

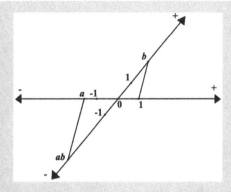

FIGURE 8.5.   The product *ab* is negative
when *a* is negative and *b* is positive.

tion point. Again, the intersection point will be the product *ab*, and again, this product is located in the negative direction.

This suggests that the product of a positive number and a negative number is a negative number.

Negative number times a negative number

Again, consider a similar construction but beginning with both *a* located in the negative direction and *b* located in the negative direction (see Figure 8.7). The construction proceeds in the same way, with *a* on the horizontal, *b* on the slanted line, and construction of the segments from the 1 location to the *b* location and then a parallel segment from the *a* location to the intersection point. Again, the intersection point will be the product *ab*, but this product is located in the positive direction.

This suggests that the product of a negative number and a negative number is a positive number.

Mathematical Focus 7

The product of two negative numbers can be shown to be positive by using properties of the real number system, including the identity and inverse properties.

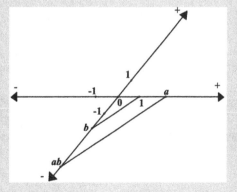

FIGURE 8.6.   The product *ab* is negative
when *a* is positive and *b* is negative.

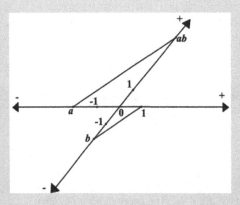

FIGURE 8.7. The product *ab* is positive
when both *a* and *b* are negative numbers.

The set of real numbers forms an ordered field under the operations of addition and multiplication as they are typically defined. In a field, additive identities and additive inverses exist. In addition, the distributive property of multiplication over addition holds. These properties can be used to show that the product of two negative numbers is positive. Negative real numbers are the additive inverses (opposites) of positive real numbers. Zero acts as the additive identity in this field. Choose any positive element from the set of real numbers and call it *a*. The following sequences of statements illustrate the use of the various properties[4] to show that the product of two negative numbers is positive.

$$a + (-a) = 0$$

$$(-1) \times (a + (-a)) = (-1) \times 0$$

$$(-1)(a) + (-1)(-a) = 0$$

$$-a + (-1)(-a) = 0$$

$$a + -a + (-1)(-a) = a + 0$$

$$(-1)(-a) = a$$

This establishes that the product of -1 and the additive inverse of any negative real number is that real number. This result is used within the following sequence of statements to show that $(-1) \cdot (-1) = 1$.

Next, let *a* and *b* be any two negative real numbers. Therefore, $a = -c = (-1)c$ and $b = -d = (-1)d$ for some positive real numbers *c* and *d*. Then

$$ab = [(-1)c] [(-1)d]$$

$= [[(-1)c](-1)]d$ by the associative property of multiplication

$= [(-1)(c(-1))]d$ by the associative property of multiplication

$= [(-1)[(-1)c]]d$ by the commutative property of multiplication

$= [[(-1)(-1)]c]d$ by the associative property of multiplication

$= (1c)d$ using $(-1)(-1) = 1$

$= cd$ by the multiplicative identity.

Thus, the product of two negative real numbers is equal to the product of two positive real numbers, which previously has been established to be positive. So the product of two negative real numbers is positive.

POSTCOMMENTARY

Having multiple ways to think about the multiplication of negative numbers is often helpful in teaching. There are many intuitively compelling reasons to have the product of two negative numbers be positive (e.g., extending the pattern generated by multiplying positive whole numbers by negatives, the result when vectors are reflected over 0 on a number line), but in all cases those amount to giving motivation for what is taken as the usual statement about the product of two negative numbers. Conceptual models and patterns such as those described in Focus 3 and Focus 5 (see also Crowley & Dunn, 1985) can be invoked to provide support for believing the result, but do not prove it.

Nonmathematical analogies[5] often are used to motivate the rule that the product of two negative numbers is a positive number. For example, one might draw an analogy between filming the filling of a glass of water and multiplication. One can film filling (positive) or emptying (negative) a glass of water, and one can run the film forward (positive) or backward (negative). Running the film backward as the glass is emptying will show it filling (suggesting that a negative times a negative equals a positive). Although such analogies are seductive because they "work"—they simply remind the user of the result (two negatives yield a positive)—they do not capture the mathematical essence of some model of multiplication (e.g., repeated accumulation of equal-sized groups). This movie analogy merely suggests that there is another situation in which two negative actions lead to a positive result. Other nonmathematical analogies include a person's action of moving into or out of a city and that person's categorization as a good person or a bad person. When two negatives occur, namely, a bad person moves out of the city, it is good for the city (a positive result). Again, this only suggests that another situation exists in which two negatives lead to a positive result.

Extension of operations and properties from one number system to an expanded number system (e.g., from integers to rational numbers or to real numbers) requires identification of a collection of assumptions (axioms) that hold for the expanded number system. Although it is tempting to apply the number properties established for one number system to another number system, it is necessary to establish the reasonableness of those properties for each new number system.

NOTES

1. Note that this Focus assumes that the behavior of vectors on the number line under multiplication by a negative scalar resembles the behavior of real numbers under multiplication by a negative number.

2. An illustration on page 31 of Hefendehl-Hebeker (1991) that the product of two negative numbers is a positive number includes the use of the additive inverse property, multiplication property of equality, and the symmetric and transitive properties of equality.

3. Note that this pattern does not suggest that the product of any two negative numbers should be positive, only that the product of -5 and a negative number should be positive. One might create tables of values for functions of the form $f(x) = kx$ for a variety of negative values of k and use the pattern across those tables as the basis for an empirical argument (but not a proof) that suggests that the product of any two negative numbers is positive.

4. Note that this sequence of statements relies on several properties having been established prior to their use here. For example, the zero-product property ($a \times 0 = 0$) is used to replace $(-1) \times 0$ by 0.

5. One can call these analogies *nonmathematical* because they do not constitute the basic elements of a mathematical argument. There are no clearly identified starting objects, or primitives, accompanied by axioms from which the mathematics community determines allowable conclusions that can be drawn or logical arguments that can be formed about the primitives based on those axioms.

REFERENCES

Crowley, M. L., & Dunn, K. A. (1985). On multiplying negative numbers. *Mathematics Teacher, 78*, 252–256.

Hefendehl-Hebeker, L. (1991). Negative numbers: Obstacles in their evolution from intuitive to intellectual constructs. *For the Learning of Mathematics, 11*(1), 26–32.

CONNECTION TO STANDARDS

Learning experiences in mathematics for teachers are well positioned when teachers are aware of connections to the standards to which their students are accountable. These standards differ across states and provinces as well as across countries, although there are commonalities across different sets of standards. Although this document cannot possibly address all the existing sets of standards across states, provinces, and countries, an example follows that illustrates how the proposed professional learning might address one particular set of standards. The example addresses the connections of the Common Core State Standards in Mathematics (CCSSM) (National Governors Association Center for Best Practices & Council of Chief State School Officers, 2010) to the proposed professional learning related to the Product of Two Negative Numbers Situation. The Common Core State Standards for mathematical content and mathematical practice mirror the attention given in various sets of standards both to what mathematics should be learned and to the mathematical processes in which students should engage. The Appendix of this Guidebook lists the CCSSM Standards for Mathematical Practice, one example of mathematical process standards. Questions (in bold italics in the following chart) that accompany the display of each CCSSM standard can be used in professional learning settings to extend work with specific standards.

Related Common Core Standards
CCSSM Standards for Mathematical Content
Grade 6 The Number System **Apply and extend previous understandings of numbers to the system of rational numbers.**
6.NS.6.a.   Recognize opposite signs of numbers as indicating locations on opposite sides of 0 on the number line; recognize that the opposite of the opposite of a number is the number itself, e.g., −(-3) = 3, and that 0 is its own opposite.   ***How is the idea of the opposite of the opposite related to the product of two negative numbers?***[3]
Grade 7 Critical Areas
Item (2).   Students develop a unified understanding of number, recognizing fractions, decimals (that have a finite or a repeating decimal representation), and percents as different representations of rational numbers. Students extend addition, subtraction, multiplication, and division to all rational numbers, maintaining the properties of operations and the relationships between addition and subtraction, and multiplication and division. By applying these properties, and by viewing negative numbers in terms of everyday contexts (e.g., amounts owed or temperatures below zero), students explain and interpret the rules for adding, subtracting, multiplying, and dividing with negative numbers. They use the arithmetic of rational numbers as they formulate expressions and equations in one variable and use these equations to solve problems.   ***What explanation and interpretation by students would be reasonable to expect regarding the "rule" for multiplying negative numbers?***

Grade 7 The Number System
Apply and extend previous understandings of operations with fractions to add, subtract, multiply, and divide rational numbers.

7.NS.2.		Apply and extend previous understandings of multiplication and division and of fractions to multiply and divide rational numbers.
	a.	Understand that multiplication is extended from fractions to rational numbers by requiring that operations continue to satisfy the properties of operations, particularly the distributive property, leading to products such as (-1)(-1) = 1 and the rules for multiplying signed numbers. Interpret products of rational numbers by describing real-world contexts.

How might the distributive property be used to establish that the product (-1)(-1) equals 1?

	b.	Understand that integers can be divided, provided that the divisor is not 0, and every quotient of integers (with nonzero divisor) is a rational number. If p and q are integers, then $-\dfrac{p}{q} = \dfrac{-p}{q} = \dfrac{p}{-q}$.

Interpret quotients of rational numbers by describing real world contexts.

How does the relationship between opposite signs and fractional representations of rational numbers connect to the product of two negative numbers? Do students confuse opposite signs and where they are located in a fraction to mean different things? What about extending this to $\dfrac{-p}{-q}$?

CCSSM Standards for Mathematical Practice[4]	
SMP2.	Reason abstractly and quantitatively.
SMP3.	Construct viable arguments and critique the reasoning of others.
SMP8.	Look for and express regularity in repeated reasoning.

SUGGESTIONS FOR USING THIS SITUATION

Facilitators may want to peruse the following chart for an idea about how to use the Product of Two Negative Numbers Situation in their professional learning settings. The chart provides an Outline of Participant Activities and a summary of the Tools and projected Time required for implementation of those activities. Following the chart are Facilitator Notes that describe each of the suggested activities in greater detail.

Tools	Time
• Poster paper, markers, or electronic means of sharing work (e.g., projection from individual laptops, shared documents online) • Copies of the Prompt (separate from the Foci) • Copies of the Foci, each beginning on a new page • Copies of the Postcommentary • Copies of Standards for Mathematical Practice (included in the Appendix)	Approximately 3–4 hours; can be done in a single session or across multiple sessions

Outline of Participant Activities (Details for these activities follow in the Facilitator Notes section.)	
Launch	Participants read the Prompt and identify instances in the secondary school curriculum in which the product of two negative numbers appears. Participants also think about (and discuss in groups) the way(s) in which one might explain to students—who already know that the sum of two negative numbers is negative—that the product of two negative numbers is positive.
Activity 1.	In small groups, participants read and discuss the Foci, eventually ranking the Foci's arguments that the product is positive according to how convincing they are to the participants.
Activity 2.	In a large group setting, participants identify how an inductive argument or an analogical argument differ from a deductive proof in the mathematics on which they are based and in the extent to which they establish a result for all possible cases.
Activity 3.	Participants share the explanations that they developed in Activity 2 and determine which group explanations from Activity 2 are intuitive arguments and which are proofs.
Activity 4.	Individually or in small groups, participants identify which of the CCSSM Standards for Mathematical Practice are addressed in the Foci and discuss those with the large group.
Reflect and assess learning	Participants identify any changes in their views of what it means to explain why a statement is true and reflect on whether they have reconsidered how they might explain "rules" to their students.

FACILITATOR NOTES

About the mathematics

Although middle school students often can explain why it is reasonable for the sum of two negative numbers to be negative, and although when faced with multiplication of negative numbers they can recite a rule such as "a negative times a negative is a positive," many of them may not understand why it is reasonable for the product of two negative numbers to be positive. The mathematical ideas in this Situation's Foci can help someone understand why a positive product is both a result that is consistent with similar (but not structurally isomorphic) settings and a necessary conclusion based on properties of the real number system.

It is important for participants to understand the difference between evidence that provides support for believing that a statement is true (e.g., a figure, examples that illustrate some pattern, an analogy) and a deductive argument that a statement is true (e.g., a proof based only on axioms that have been assumed to be true, definitions, and previously proven theorems). Be sure that participants recognize that the mathematics in Focus 1, Focus 2, Focus 3, Focus 5, and Focus 6 of this Situation only suggest that the product of two negative numbers is positive, whereas the mathematics in other Foci (Focus 4 and Focus 7) is quite different because it offers a deductive proof—drawing on properties of the real number system—that the product of two negative numbers is positive.

Launch

Time

30–35 minutes

Conducting the Launch

In the two-part Launch, emphasize that "knowing the rule" is not enough. Teachers need not only to understand why the product of two negative numbers is positive but also to be able to offer to students convincing arguments that the product should be positive. Begin by asking participants where this idea occurs in the mathematics curriculum. The chart in the Connections to Standards section identifies where the product of two negative numbers occurs in the Common Core State Standards for Mathematics (National Governors Association Center for Best Practices and Council of Chief State School Officers, 2010). Continue with the main portion of the launch, namely, asking participants how they might explain why the product of two negative numbers is positive.

Launch Part 1. Read the Prompt and identify instances in secondary school curricula in which the product of two negative numbers appears.

After participants read the Prompt, have them suggest instances in their mathematics curriculum in which the product of two negative numbers is addressed or applied. Also have participants discuss how well they believe that their students understand this property and what contributes to students' misunderstandings.

Anticipated participant responses

Participants are likely to note that in middle school curricula the product of negative numbers arises as part of the study of operations on integers. In high school, symbolic manipulation of algebraic expressions often requires application of the rule for the product of two negative numbers. In algebraic expressions, sometimes the two factors being multiplied are negative integers, as is the case with the constant term that results from expanding a product of binomials such as $(c - 6)$ $(c - 5)$. Sometimes the product might include a negative integer and a monomial with a negative coefficient, for example, $(-4)(-3x)$. In other instances, each of the two factors might be the opposite of a polynomial expression (but not necessarily a negative number), for example, $[-(3a + 5)] \times [-(a + 2)]$.

Facilitating Launch Part 1

If the participants are all middle school teachers or all high school teachers, be sure that each group identifies instances in both middle school and high school curricula in which students encounter the product of two negative numbers. (Have the participants focus on the larger instances in the curriculum in which factors involving negative numbers occur, e.g., operations on integers or use of the distributive property, rather than on individual problems that might require calculation of the product of two negative numbers.)

Have a few groups report the instances they discussed, so that one or more middle school topics are presented and one or more high school topics are presented. Refer to the CCSSM Content Standards for Grades 6 and 7 that are identified in the preceding Connections to Standards section and address some of the accompanying questions as appropriate for the participant audience.

Launch Part 2. Think about (and discuss in groups) the way(s) in which one might explain to students— who already know that the sum of two negative numbers is negative—that the product of two negative numbers is positive.

Next, have participants individually address the Prompt by thinking of an explanation (for students who already know that the sum of two negative numbers is negative) for why the product of two negative numbers is a positive number. Participants might focus their students especially on thinking about why addition of two negative numbers results in a negative number but multiplication of two negative numbers results in a positive number. Participants should share their explanations with their partners or small-group members. If participants produce only one explanation, encourage them to develop another explanation.

Anticipated participant responses

Participants might develop an explanation based on interpretation of multiplication as repeated addition (as in Focus 1). They might also justify empirically that the product of two negative numbers is positive by considering sequences of products and basing their conclusion on a pattern in those examples (as in Focus 5). They might also rely on real-world settings similar to that in Focus 2 that suggest analogies with multiplication of negative numbers. Some participants might employ one or more (essentially nonmathematical) analogies such as:

Multiplying negative numbers is like desirable and undesirable people moving into or out of a town. Say that moving in is positive and moving out is negative. Similarly, a desirable person is positive and an undesirable person is negative. The product involves a person (desirable/undesirable) with a particular sign and a direction (in/out) with a particular sign. So, when an undesirable person (negative person) moves out of town (negative direction), the result is positive, because when the town loses population because an undesirable person leaves, it is good for the town.

Participants might offer an intuitive argument similar to that presented at the beginning of Focus 4 or the one in Hefendehl-Hebeker (1991):

Given the equations (*) and (**),

$$-3 + 3 = 0 \ (*)$$

$$-4 + 4 = 0 \ (**),$$

multiply both members of equation (*) by 4 and both members of equation (**) by -3, yielding

$$(-3)(4) + (3)(4) = 0 \ (*)$$

$$(-3)(-4) + (-3)(4) = 0 \ (**).$$

The left members are both equal to 0, so equating the left members yields

$$(-3)(-4) + (-3)(4) = (-3)(4) + (3)(4). \ (***)$$

Subtracting (-3)(4) from both members of (***) leaves (-3)(-4) = 12, suggesting that the product of (these) two negative numbers is a positive number.

Some participants might develop a deductive algebraic proof that the product of two negative numbers is positive, whereas others might base their explanations on a model developed with manipulatives such as Algeblocks.

Facilitating Launch Part 2

Have participants discuss the arguments generated within their groups and record their work on poster paper or electronically, but do not have groups report out. Some groups' responses may be inductive arguments that suggest—but do not prove—that the product of two negative numbers is positive, and others may be deductive proofs. Carefully note the groups that develop compelling arguments that only suggest that the result is true and any that develop a deductive proof. In a subsequent activity, groups will present their arguments and participants will be asked to judge whether other groups' arguments are proofs or simply justifications that suggest that the product should be positive. Some facilitator questions that might help to foreshadow that discussion might be ones such as "How convincing do you think your explanation would be?" "Do you think that your explanation would convince a student that the product of any two negative numbers is positive?"

Activity 1. In small groups, have participants read and discuss the Foci, eventually ranking the arguments in them according to how convincing they are that the product is positive.

Read and discuss each of the Foci, ranking the Foci according to which presents the most compelling argument for the product being positive. Then note which are intuitive arguments (those in Focus 1, Focus 2, Focus 3, and Focus 5) and which are proofs (e.g., the ones in Focus 4 and Focus 7).

Time

Focus 1: 20 minutes

Focus 2: 10 minutes

Focus 3: 15 minutes

Focus 4: 25 minutes

Focus 5: 15 minutes

Focus 6: 30 minutes

Focus 7: 30 minutes

Anticipated participant responses

Participants may rank the arguments in Focus 1 as being more convincing than the others, given that it is based on a pattern that students should be able to notice easily.

Key points about the Foci

The model in Focus 1, repeated addition, is likely to be the most common model that teachers (and their students) are likely to have for multiplication. This model, which is a straightforward one when applied to positive integer multiplication, is not as straightforward when used with negative integer multiplication. It is important for participants to consider the types of negative number factors for which repeated addition offers a straightforward model for the product and those negative number factors for which it does not. Although one can easily think of 3 groups of -5 being added together, it is more difficult to think of what is meant by -3 groups of -5 being added together. Table 8.1 in the Product of Two Negative Numbers Situation illustrates how one might use the pattern that results when one decreases the number of negative addends from a positive number, through 0, to a "negative number of negative addends." Each decrease of one in the number of addends results in removal of one negative addend, causing the sum to increase by 5. Although this model suggests that for some pairs of negative number factors the product is positive, an even greater limitation of this model is its inability to convey what is meant when neither of the negative factors is an integer. Repeated addition can suggest that, in some instances, the product of two negative numbers is a positive number, but it does not provide the basis for a proof that the product ab is positive for all $a, b < 0$.

Teachers often rely on analogies to justify mathematical results. Focus 2 and Focus 3 offer examples of analogies: a real-world one in Focus 2 and a vector-based one in Focus 3. Focus 2 draws an analogy between multiplication of negative numbers and the removal of an amount debited from an employee's paycheck over a sequence of months. The result suggests that the product of two negative numbers is positive.

Focus 3 draws on knowledge of the results of scalar multiplication of a vector in one dimension by the scalar -1. That multiplication results in the original vector being reflected about the origin. The vector $-(-j)$ corresponds to the vector j because the composition of two consecutive reflections about the origin results in the image being the original vector j. This suggests that $(-1)(-1) = 1$. Again, the argument offered in Focus 3 for the product being positive is an intuitive one that is based on an analogy between multiplication of negative numbers and scalar multiplication of vectors. As such, it does not constitute a proof that the product of two negative numbers is a positive number.

Focus 4 utilizes the distributive property of multiplication over addition both in specific examples in which two negative factors produce a positive product and in a general argument that the product of two negative numbers will always be positive. Use of properties of the real number system enables one to generate a general argument that holds for all pairs of negative real number factors.

The patterns in the tables in Focus 5 should be apparent and easily understood. What is important for participants to understand is that the entries in the rightmost column of Table 8.2 in the Product of Two Negative Numbers Situation are generated from two known results: The product of a negative number and a positive number is negative, and the product of 0 and any number is 0. However, the entries in the rightmost column of Table 8.3 in the Product of Two Negative Numbers Situation are not generated from known results but from continuation of the pattern that is apparent in Table 8.2 in the Product of Two Negative Numbers Situation. Thus, there is an assumption that the function *f* behaves in some regular fashion—with increases of 5 in *f*(*x*) for each decrease of 1 in *x*. The entries in the rightmost column of Table 8.3 in the Product of Two Negative Numbers Situation can be justified—as noted in Focus 5—by examining the rate of change of the linear function *f*, for which a change of 1 in *x* corresponds to a change of -5 in *y*. The results in the table thus are suggestive of what occurs when other negative numbers are multiplied by -5. If the observed pattern continues, that pattern suggests—but does not prove—that the product of -5 and other negative numbers will be a positive number. Another key point in Focus 5 is addressed in Note 3 of the Situation: These tables suggest results for the product of -5 and a negative number, but they do not establish what will happen when a negative number is multiplied by a negative number other than -5. These "shortcomings" of reliance on patterns in tables of examples underscore the need for a more general, deductive proof that the product of two negative numbers is a positive number.

Focus 6 presents a geometric argument—based on similar triangles—that the product of two negative numbers is a positive number.

In Focus 7, a proof is presented for the product of two negative numbers being a positive number. That proof is based on the properties of the real number system. Participants will need to be familiar with (or become familiar with) the concepts of opposite, additive inverse, and multiplicative identity as well as with the associative and commutative properties of multiplication and the distributive property of multiplication over addition. In contrast to the arguments presented in Focus 1, Focus 2, Focus 3, and Focus 5, the argument presented in Focus 7 is a proof.

Facilitating the activity

With Focus 1, raise the question of whether repeated addition (which is a common model for multiplication of positive numbers) is applicable to the product of one negative factor and one positive factor or the product of two negative factors. After participants read Focus 1, have them address the issue of how one might interpret a negative number of addends when repeated addition is used with products of two negative factors. Also have them discuss the limitation inherent in applying repeated addition with products of two nonintegral negative factors. If participants do not raise the issue, one might ask, "Is the repeated-addition model for multiplication applicable when both factors are negative numbers?" "How, if at all, does the model apply if one of the factors is not an integer?" "How, if at all, does the model apply if both factors are not integers?" "How, if at all, does the model apply if both factors are irrational numbers?"

The application involving debiting and crediting an employee's paycheck in Focus 2 is one such real-world application that suggests a positive product when two negative factors are multiplied. Have participants generate other real-world settings (e.g., ones that involve forgiving debts or reversing withdrawals from an account) that might be used to craft similar intuitive arguments about the sign of the product of two negative numbers. Participants might be asked how their setting relates to multiplication and whether their setting is one that is analogous to multiplication or is merely a setting that involves an analogous result (two negatives produce a positive). One also might ask, "Does your setting lead to an argument that proves that the product of any two negative numbers is a positive number?"

In Focus 3, the topic of vectors may not be as familiar to middle school teachers as it might be to some high school teachers, so some additional explanation might be necessary, particularly if the participants are primarily middle school teachers who have been certified through an elementary teacher preparation program rather than a secondary mathematics teacher preparation program. The figures illustrating scalar multiplication of one-dimensional vectors offer arguments for the product of two negative numbers being a positive number that rest on an analogy that multiplication by a negative scalar is "like" multiplication by a negative number, and thus, similar "rules" should hold. Have participants read Focus 3 and, afterward, have them describe how multiplication of a negative vector by a negative scalar and multiplication of a negative number by another negative number are analogous.

One way to approach Focus 4 might be to have participants read the first two paragraphs that address the products $(-4) \times (-2)$ and $(-2) \times (-3)$, noting how properties of the real number system are used to show that the product of a particular pair of negative factors is a positive number. Before going on to the generalized argument that follows, ask participants to generate similar instances in which the distributive property enables one to show that the product of a particular pair of negative factors is a positive number. Subsequently, they should examine the generalized argument, noting where the distributive property is used (in factoring) and how this argument differs from those that address particular numerical products. After completing Focus 4, participants should be able to generate similar specific or general arguments.

Have participants read Focus 5, and ask them two questions: Are they convinced by the explanation presented, and if so, of what are they convinced? Some participants may view the Focus as having shown that the product of any two negative numbers will be a positive number; in actuality, Focus 5 only suggests that -5 times a negative integer will be a positive number. Also question participants about the origin of the values in the rightmost columns of Table 8.2 and Table 8.3 in the Product of Two Negative Numbers Situation. Ask whether they are derived from known properties or generated inductively based on a collection of examples.

Focus 6 presents a geometric interpretation of multiplication of negative numbers. It involves locating the values corresponding to two positive, one positive and one negative, and two negative factors at points on the positive and/or negative portions of two intersecting number lines. Those intersecting lines constitute an *oblique coordinate system*. Participants may need help in interpreting the figures in Focus 6. For example, in Figure 8.4 of the Situation, consider finding the length of the segment from the point a units to the right of 0 on the horizontal line to the point labeled as being a distance of ab from 0 on the slanted line. The proportion $1:b = a:x$ based on similar triangles leads to x being equivalent to ab, providing corroboration that the point labeled as being a distance of ab from 0 on the slanted line actually has the indicated length, ab. For each of Figures 8.4–8.7 of the Situation, have participants identify a proportion that results from the similar triangles in the figure.

Have students read Focus 7 and determine the properties that justify each of the steps in the demonstration that $(-1)(-a) = a$. (Properties are given to justify each step of the more general demonstration that the product of two negative real numbers, ab, is equal to the product of two positive real numbers, cd.) Depending on participants' mathematics backgrounds, different amounts of explanation might be necessary as participants work through Focus 7.

Once participants have reviewed all the Foci, in small groups have them rank the explanations in the Foci according to how convincing they are to them. Such rankings often engender good discussion, particularly if individuals' rankings differ from those of other members of their group, or if different groups' rankings vary greatly. Also discuss how convincing the seven Foci would be to their students. If it has not arisen already, address whether the arguments in each of the Foci are intuitive arguments that suggest that the product of two negative numbers is a positive number or whether they are proofs of that claim.

Activity 2. With the entire group, identify how an intuitive argument and a deductive proof differ in the mathematics on which they are based and the extent to which they establish a result for all possible cases.

Based on the nature of the arguments presented in the seven Foci, in small groups, have participants describe the difference(s) between an intuitive argument that suggests that a statement is true and a deductive proof that it is true. In the large group, discuss the nature of the arguments in each of the seven Foci and the certainty with which they establish the rule for multiplying two negative numbers.

Time

10–15 minutes

Anticipated participant responses

Participants might describe intuitive arguments based on surface features such as their use of pictures or examples and proofs in terms of surface features such as their format (e.g., two-column proofs that list reasons for each step). Help the participants to focus instead on the basis of the arguments and the extent to which the arguments establish the truth of a statement for all possible instances. For example, Focus 5 is based on the pattern in a set of examples and establishes the result for only a limited number of cases. An analogy forms the basis for the arguments in Focus 2 and Focus 3, and although those analogies might be considered to be more general than the arguments in Focus 1 and Focus 5, they are only intuitive arguments, not ones that establish truth for all cases. In contrast, the deductive arguments in Focus 4 and Focus 7 are based on accepted assumptions about the real number system and previously proven results and establish the result for all possible pairs of negative real number factors.

Facilitating the activity

Have small groups record on poster paper or via electronic device their description of the differences between an intuitive argument and a deductive proof. A two-column table with headings *Intuitive Argument* and *Deductive Proof* might serve as a way for groups to organize their comparisons. What they produce can be shared with and critiqued by the large group.

After discussing the nature of the arguments in each of the Foci, it might be helpful for participants to read the Post-commentary, which conveys two important ideas. First, even with a deductive proof, results are based on a set of assumptions. Second, some analogies simply remind one of the result (two negatives yield a positive), but they do not embody the mathematical essence of multiplication.

Activity 3. Share the explanations developed in Launch Part 2 and determine which groups' explanations from Launch Part 2 are intuitive arguments and which are proofs.

Small groups should share with the large group their explanations developed in Launch Part 2, and the large group should determine which are intuitive arguments and which are proofs.

Time

40–60 minutes

Anticipated participant responses

Participants will present what they recorded in Launch Part 2 for discussion by the large group. Participants' explanations from Launch Part 2 may fall into three categories: (a) deductive proofs (as in Focus 4 and Focus 7 in the Product of Two Negative Numbers Situation), (b) nonmathematical analogies that are based on real-life situations that have physical characteristics that are analogous to the mathematical characteristics of the product of two negative numbers (e.g., filming the filling and emptying of a glass of water and running the film forward and backward), and (c) nonmathematical analogies with real-life analogues whose analogous results are not based on mathematical relationships (e.g., good and bad persons moving in and out of a city). (See the Postcommentary of the Product of Two Negative Numbers Situation for a discussion of the examples in (b) and (c).) Participants may tend to have difficulty differentiating among the three categories because they may not have considered the nature of the analogies that often are used in classrooms.

Facilitating the activity

Have a selection of small groups of participants present the explanation(s) they generated within their group for Launch Part 2 and have the other participants decide whether they have presented an intuitive argument or a deductive proof. As facilitator, you may have noted—as suggested in Launch Part 2—which groups developed compelling arguments that only suggested that the result was true and which groups developed a deductive proof that established that the product of any two negative numbers is positive. You might use that information to select the groups that are asked to present their explanation(s). If there is time and if groups' explanations differ substantially, all small groups might present their explanation(s). Include questions about which mathematical practice(s) they are using in their explanation, keeping in mind that their responses do not need to focus on the subset of practices listed in this chapter.

Ask the large group whether each explanation presented merely suggests that the product of two negative numbers is positive or proves that the product of two negative numbers is a positive number. Include whether such questions have ever come up, or might come up, in their classrooms and how they have or would address them with students.

Key points for discussion

- Does this explanation convince you that the product of any two negative numbers is a positive number? Would it convince your students?

- On what is the explanation based (e.g., examples, a mathematical analogy, a picture, a nonmathematical analogy, properties of the real number system)?

- Is the explanation an intuitive argument that suggests that the product of two negative numbers is a positive number or a deductive sequence of statements that proves that the product of any two negative numbers is a positive number?

Activity 4. Identify individually or in small groups which of the CCSSM Standards for Mathematical Practices are addressed in the Foci and discuss those in the large group.

Time

15 minutes

Anticipated participant responses

Participants are likely to identify SMP 3 (critiquing the reasoning of others) and SMP 8 (recognizing regularity in repeated reasoning) as mathematical practices that might be addressed when students encounter the product of two negative numbers. They might be less likely to note that SMP 2 might be addressed (addressing the meaning of quantities, not simply how to compute them).

Facilitating the activity

Ask participants to note the Standards for Mathematical Practice in which they were engaged during the session (see the Appendix to this Guidebook). Include the practices in which they were engaged as they (a) analyzed the Foci, and (b) critiqued other groups' presentations.

Reflect and assess learning

The purpose of the reflection is for participants to assess what they learned from the session and to connect the session's activities to their own classroom practice. Specific activities depend on the setting for the professional learning session, that is, whether it is a stand-alone session or part of an ongoing series of sessions.

Time

30–60 minutes

Suggested assessment activities

- Ask participants to reflect on whether and, if so, how their views of what it means to explain why a statement is true may have changed during their work on this Situation.

- Ask participants to reflect on how, if at all, they have reconsidered how they might explain "rules" to their students.

- Ask participants to list any unanswered questions that remain as a result of their work on this Situation.

- If this session is part of an ongoing series of professional learning sessions, participants could be asked to examine explanations that textbooks use for why the product of two negative numbers is a positive number, classifying them as either intuitive arguments or deductive proofs.

- Similarly, if this session is part of an ongoing series of sessions, participants can be asked to select some topic in their curriculum that requires explanation of some often-used rule, develop an explanation that is targeted to students, and then bring classroom artifacts to the next session that might convey the extent to which students understood the explanation in which the teacher engaged students. Such artifacts might include, but not be limited to, video of the classroom session, journal entries from students who have given permission for their work to be shared, a collection of students' written explanations of why a particular statement is true, or students' attempts to prove a particular statement. If participants include several teachers from a given school, those participants might observe each other's implementations of their explanations, offering critique after class and support during the next professional learning session when the teacher explains her or his implementation of the explanation with students.

Reflection questions

1. How, if at all, has our work with the seven Foci caused you to reconsider what it means to explain why a statement is true? Explain.

2. Has our work with this Situation caused you to reconsider the way(s) in which you explain "rules" to students? Explain.

3. How, if at all, has our work with this Situation engaged us in mathematical practices? Explain.

4. How might our work with this Situation help us to use representations other than the vector, algebraic, and geometric representations presented in this Situation to investigate operations on various types of numbers?

5. Are there unanswered questions that your work with this Situation has raised?

NOTES

1. The Product of Two Negative Numbers Situation is one of the Situations presented in *Mathematical Understanding for Secondary Teaching: A Framework and Classroom-Based Situations* (Heid & Wilson, 2015).

2. This Situation appears on pp. 103–115 of Heid and Wilson (2015). It is reprinted with permission.

3. For each standard, we have suggested questions (those in italicized, boldface type) that might serve as follow-up questions to address the mathematics of the standard.

4. See the Appendix for the list of the Standards for Mathematical Practice in the Common Core State Standards for Mathematics (National Governors Association Center for Best Practices and Council of Chief State School Officers, 2010).

REFERENCES

Hefendehl-Hebeker, L. (1991). Negative numbers: Obstacles in their evolution from intuitive to intellectual constructs. *For the Learning of Mathematics, 11*(1), 26–32.

Heid, M. K. & Wilson, P. W. (with G. W. Blume). (Eds.). (2015). *Mathematical understanding for secondary teaching: A framework and classroom-based situations*. Charlotte, NC: Information Age.

National Governors Association Center for Best Practices & Council of Chief State School Officers. (2010). *Common core state standards for mathematics*. Washington, DC: Authors. Retrieved from www.corestandards.org/uploads/Math_Standards1.pdf

CHAPTER 4

FACILITATOR'S GUIDE
FOR
GRAPHING QUADRATIC FUNCTIONS

Situation 21 From the MACMTL–CPTM Situations Project[1]

M. Kathleen Heid

A secondary mathematics curriculum contains many formulas. Rather than simply presenting students with a formula as something to remember followed by practice in using the formula to compute values for particular quantities, teachers might ask how the formula connects to other things that students know. One way in which teachers might make sense of the formula or share it with students is through varied derivations of a formula. This Situation engages teachers in deriving the equation for the axis of symmetry of a parabola using several different tools, including different types of representations and different areas of secondary school mathematics.

Facilitator's Guidebook for Use of Mathematics Situations in Professional Learning,
pages 59–83.

OVERVIEW

Facilitators can scan the following overview to quickly get a sense of the mathematics involved in the proposed professional learning setting. The sections deal with the mathematics of the Graphing Quadratic Functions Situation, why these mathematical ideas might be important for participants, the learning goals for participants, and mathematical ideas central to the proposed professional learning sessions.

Situation	Relevance
This Situation addresses different derivations for the equation, $x = \dfrac{-b}{2a}$, for the axis of symmetry for a parabola with function rule $y = ax^2 + bx + c$, $a \neq 0$.	• The derivation of the equation for the axis of symmetry provides an example of approaching an important mathematical result from geometric, calculus-based, and transformational approaches. • Teachers often know of or focus on only one way to justify a result. Teachers can use this as a way to engage students in developing both graphical and symbolic analyses of the same problem.
Goals	Key mathematical ideas
• Increase teachers' understanding of the derivation of the equation for the axis of symmetry for a quadratic function. • Expand teachers' ability to reason from the definition and properties of a quadratic function to apply geometric, calculus, and transformational approaches to develop the equation for the axis of symmetry. • Increase teachers' capacity to articulate an approach and reason through different areas of mathematics and with different representational forms. • Enhance teachers' ability to analyze prerequisite understandings on which each of the different approaches to an explanation could be built.	• For any point on the graph of a quadratic function other than the vertex, there is another point on the graph that is equidistant from the line of symmetry. • In a plane, a parabola is the set of points each of which is equidistant from a given line and a given point (i.e., locus definition of parabola). • The vertex is the point of the graph of a quadratic function representing the function's minimum or maximum value. • Algebraically, vertical and horizontal stretches, contractions, and translations of the graph of a function, f, correspond to altering values of h_1, h_2, v_1, or v_2 given $g(x) = v_1 f(h_2(x - h_1)) + v_2$. • Viète's formulas for the relationship between r_1 and r_2, the roots of the function $y = ax^2 + bx + c$, $a \neq 0$, and the coefficients a, b, and c are $r_1 + r_2 = -\dfrac{b}{a}$ and $r_1 r_2 = \dfrac{c}{a}$.

The Situation under consideration in this chapter follows (highlighted with a gray background).[2]

GRAPHING QUADRATIC FUNCTIONS

Situation 21 From the MACMTL–CPTM Situations Project

Ginger Rhodes, Ryan Fox, Shiv Karunakaran, Rose Mary Zbiek, Brian Gleason, and Shawn Broderick

PROMPT

When preparing a lesson on graphing quadratic functions, a student teacher found that the textbook for the class claimed that $x = \dfrac{-b}{2a}$ was the equation for the line of symmetry of a parabola $y = ax^2 + bx + c$. The student teacher wondered how this equation was derived.

COMMENTARY

This Prompt addresses graphing quadratic equations, specifically the derivation of the equation of the line of symmetry of a parabola. The Foci in this Situation deal with the general symbolic representation of a quadratic function, but they differ in the approaches used to obtain the equation in question. Focus 1 uses the symmetry of the parabola to find the x-coordinate of the vertex of the parabola. Focus 2 uses the first derivative to find the x-coordinate of the vertex of the parabola. Focus 3 utilizes transformations of the graph of $y = x^2$ to determine the coordinates of the vertex. Focus 4 uses some results about the roots of a polynomial equation, generally known as Viète's formulas, to find the x-coordinate of the vertex of the parabola.

MATHEMATICAL FOCI

Mathematical Focus 1

Knowing the general form of the equation representing the graph of a parabola can enable one to identify the line of symmetry.

A quadratic function can be written in the general form $y = ax^2 + bx + c$, where a, b, and c are real numbers and $a \neq 0$. The graph of this function is a parabola, for which the line of symmetry is the line through the focus that is perpendicular to the directrix. This parabola is symmetric about a line $x = k$, because its directrix is parallel to the x-axis. The symmetry of the graph about the line $x = k$ implies that, for example, the function values $f(k-1)$ and $f(k+1)$ must be equal:

$$a(k-1)^2 + b(k-1) + c = a(k+1)^2 + b(k+1) + c.$$

Expanding the expressions on both sides of the preceding equation yields:

$$ak^2 - 2ak + a + bk - b + c = ak^2 + 2ak + a + bk + b + c.$$

Combining like terms, the equation simplifies to:

$$-2ak - b = 2ak + b.$$

Solving for k reveals that:

$$k = -\frac{b}{2a}$$

Mathematical Focus 2

The first derivative of a polynomial function can be used to obtain the coordinates of the relative extrema of the function. In a parabola, this corresponds to the vertex, the x-coordinate of which gives the x-coordinate of all points on the line of symmetry.

Polynomial functions are differentiable, and one can use the first derivative of the polynomial function to obtain its critical values. These critical values enable one to obtain the coordinates of the vertex of the parabola (the absolute minimum or the absolute maximum).

$$y = ax^2 + bx + c$$
$$y' = 2ax + b$$

Finding the critical values for a quadratic function,[1] and thus the vertex and the equation of the line of symmetry, involves setting the derivative equal to 0 and solving for x:

$$2ax + b = 0$$
$$x = -\frac{b}{2a}.$$

This critical value is the x-coordinate of the vertex. Thus, the equation of the line of symmetry will be of the form $x = -\frac{b}{2a}$.

Mathematical Focus 3

Using transformations, the graph of the function $y = x^2$ *can be mapped to the graph of any quadratic function of the form* $y = ax^2 + bx + c$. *The graph of the function given by* $g(x) = v_1 f(h_2(x - h_1)) + v_2$ *is the image of the graph of f under the composition of a horizontal translation, a horizontal stretch or contraction, a vertical stretch or contraction, and a vertical translation related to the values of* $h_1, h_2, v_1,$ *and* $v_2,$ *respectively.*

The point $(0, 0)$ is the vertex of the graph of the function given by $y = x^2$. To find the coordinates of the vertex of the general parabola, the method known as *completing the square* can be applied to the general form of the equation of a parabola, which yields a form given by the transformations noted previously.

$$y = ax^2 + bx + c$$

$$y - c = ax^2 + bx$$

$$\frac{1}{a}(y - c) = x^2 + \frac{b}{a}x$$

$$\frac{1}{a}(y - c) + \left(\frac{b}{2a}\right)^2 = x^2 + \frac{b}{a}x + \left(\frac{b}{2a}\right)^2$$

$$\frac{1}{a}\left(y - c + \frac{b^2}{4a}\right) = \left(x + \frac{b}{2a}\right)^2$$

Solving for y yields a symbolic form of the equation that can be compared to $y = x^2$, the equation of the particular case, to get information about the needed transformation:

$$y = a\left(x + \frac{b}{2a}\right)^2 + \left(c - \frac{b^2}{4a}\right)$$

The presence of $\left(x + \frac{b}{2a}\right)^2$ rather than x^2 implies a horizontal translation through $\frac{b}{2a}$ units in the negative direction (see Situation 25: Translation of Functions, in Chapter 31). This horizontal translation of the graph maps the vertex of the parabola from $(0, 0)$ to $\left(\frac{-b}{2a}, 0\right)$. The rest of the equation suggests a vertical stretch or contraction (by a factor of a) and a vertical translation (through $c - \frac{b^2}{4a}$ units in the positive direction), neither of which affect the axis of symmetry. Thus, the line of symmetry that passes through the vertex of the graph of the general parabola has the equation $x = -\frac{b}{2a}$.

Mathematical Focus 4

General facts about the roots of polynomial equations, known as Viète's formulas, can quickly yield information about the line of symmetry of a parabola.[2]

Given the roots of a quadratic function, r_1 and r_2, a result of François Viète states that $r_1 + r_2 = -\dfrac{b}{a}$ (and $r_1 r_2 = \dfrac{c}{a}$).[3] Because the roots of a quadratic are symmetric about the axis of symmetry (which contains the vertex), $\dfrac{r_1 + r_2}{2} = -\dfrac{b}{2a}$ is the x-coordinate of the vertex. Therefore, $x = -\dfrac{b}{2a}$ is the equation of the line of symmetry of the parabola.

NOTES

1. For polynomial functions in general, a critical value is tested to see whether a point is an extremum rather than an inflection point. For a parabola, the result is evident (either an absolute maximum or an absolute minimum).

2. Viète's formulas also give a way to show that two possible solutions, r and s, satisfy the equation $ax^2 + bx + c = 0$. One need only show that $r + s = -\dfrac{b}{a}$ and $r \cdot s = \dfrac{c}{a}$.

For further discussion of Viète's formula, see http://mathworld.wolfram.com/VietasFormulas.html.

CONNECTION TO STANDARDS

Learning experiences in mathematics for teachers are well positioned when teachers are aware of connections to the standards for which their students are accountable. These standards differ across states and provinces as well as across countries, although there are commonalities across different sets of standards. Although this document cannot possibly address all the existing sets of standards across states, provinces, and countries, an example follows that illustrates how the proposed professional learning might address one particular set of standards. The example addresses the connections of the Common Core State Standards for Mathematics (CCSSM) (National Governors Association Center for Best Practices & Council of Chief State School Officers, 2010) to the proposed professional learning related to the Graphing Quadratic Functions Situation. The Common Core State Standards for mathematical content and mathematical practice mirror the attention given in various sets of standards both to what mathematics should be learned and to the mathematical processes in which students should engage. The Appendix of this Guidebook lists the CCSSM Standards for Mathematical Practice, one example of mathematical process standards. Questions (in bold italics in the following chart) that accompany the display of each CCSSM standard can be used in professional learning settings to extend work with specific standards.

Related Common Core Standards
CCSSM Standards for Mathematical Content
Grade 4 Geometry   **Draw and identify lines and angles, and classify shapes by properties of their lines and angles.**   **4.G.3.**   Recognize a line of symmetry for a two-dimensional figure as a line across the figure such that the figure can be folded along the line into matching parts. Identify line-symmetric figures and draw lines of symmetry.   ***Can a parabola have more than one line of symmetry?***[3]   ***Which conic sections have a line or lines of symmetry? For each conic section that is symmetric about a line or lines, identify the line(s) of symmetry.***
High School. Functions   **Interpreting Functions**   **F.IF.2.**   Use function notation, evaluate functions for inputs in their domains, and interpret statements that use function notation in terms of a context.   ***The height of a ball thrown into the air from the top of a 50-meter tower with an initial vertical velocity of 40 meters/second is described by the function rule*** $h(t) = -4.9t^2 + 40t + 50$***. Why does is make sense that*** $h(0) = 50$***? How does the term*** $= -4.9t^2$ ***affect the height of the ball?***   **F-IF.7.**   Graph functions expressed symbolically and show key features of the graph, by hand in simple cases and using technology for more complicated cases.   a.   Graph linear and quadratic functions and show intercepts, maxima, and minima.   ***How are intercepts, maxima, and/or minima identifiable from the graphical representations and from the symbolic representations of linear functions and quadratic functions?***   ***How do convention and context aid in determining whether the values that we read from graphs should be treated as exact values or as approximations?***

F.IF.8.	Write a function defined by an expression in different but equivalent forms to reveal and explain different properties of the function.
	a. Use the process of factoring and completing the square in a quadratic function to show zeros, extreme values, and symmetry of the graph, and interpret these in terms of a context.
	How does the function rule $f(x) = a(x - b)^2 + c$ *show the symmetry of the graph of the function* **f**?

High School. Functions
Building Functions

F.BF.1.	Write a function that describes a relationship between two quantities.
	c. (+) Compose functions.
	How could the function $f(x) = a(x - b)^2 + c$ *be expressed by two different compositions of functions?*
F.BF.3.	Identify the effect on the graph of replacing $f(x)$ by $f(x) + k$, $k\,f(x)$, $f(kx)$, and $f(x + k)$ for specific values of k (both positive and negative); find the value of k when given the graphs. Experiment with cases and illustrate an explanation of the effects on the graph using technology.
	What compositions of transformations (e.g., specific translations, reflections, dilations) could be used to generate the function $f(x) = a(x - b)^2 + c$ *from the function* $g(x) = x^2$?

High School. Functions
Trigonometric Functions

F.TF.4. (+)[4]	Use the unit circle to explain symmetry (odd and even) and periodicity of trigonometric functions.
	What is the relationship between a function being odd or even and its symmetry?

CCSSM Standards for Mathematical Practice[5]

SMP2. Reason abstractly and quantitatively.
SMP3. Construct viable arguments and critique the reasoning of others.
SMP7. Look for and make use of structure.

SUGGESTIONS FOR USING THIS SITUATION

Facilitators may want to peruse what follows for an idea about how to use the Graphing Quadratic Functions Situation in their professional learning settings. The chart provides an Outline of Participant Activities and a summary of the Tools and projected Time required for implementation of those activities. Following the outline are Facilitator Notes that describe each of the suggested activities in greater detail.

Tools	Time
• Newsprint or other large space that can be used to display responses throughout the session • Technology to facilitate projection from individuals' laptops of access to online space using something such as Dropbox or GoogleDrive to facilitate electronic sharing of participant responses • Computer algebra systems (CAS) for each group of participants; ideally, participants would have individual CAS access. • Handouts with the Prompt, and the four Foci each on a separate page	3.5 hours

Outline of Participant Activities (Details for these activities follow in the Facilitator Notes section.)	
Launch	Participants recall the locus definition of a parabola and create a dynamic geometry construction that illustrates that definition.
Activity 1.	Participants read the Prompt and make sense of the formula for the axis of symmetry of a parabola.
Activity 2.	In groups, participants discuss the assumptions that are made in each Focus.
Activity 3.	Participants read the Foci, and in groups, identify the mathematical background that the student teacher in the Prompt might draw upon or that a student might need for each of the Foci to be appropriate for him or her.
Activity 4.	Groups of participants investigate how each type of transformation in Focus 3 (horizontal or vertical translation, horizontal or vertical stretch, horizontal or vertical contraction) affects the graph of the function.
Activity 5.	As time allows, groups of participants can consider how the content of the Foci might be used to connect with other topics in the secondary school curriculum. For example, how do Viète's formulas for the roots of a polynomial relate to techniques that students use to factor quadratic expressions?
Reflect and assess learning	Participants reflect on how some versions of the activities might be used in their classrooms.

FACILITATOR NOTES

About the mathematics

Participants are likely to be familiar with graphing quadratic functions and perhaps aware of the equations for the axis of symmetry but not necessarily acquainted with the approaches to analyze the formula that are used in these mathematical Foci. Attention may need to be drawn to using properties of quadratic functions and parabolas (e.g., symmetry, relationship of roots to coefficients of the function expression written in standard form) to derive the equation of the line of symmetry.

A transformation approach such as the one in Focus 3 may not be in participants' repertoires. If so, they may need a quick session on the symbolic representation of translations, stretches, and contractions.

Some of the ideas used in these Foci can be extended to other functions. Focus 1 demonstrates the use of vertical lines of symmetry. Focus 3 demonstrates an approach to analyzing functions that relies on transforming parent functions. Focus 4 applies Viète's formula to quadratic functions, but the general formula is descriptive of polynomial functions of any finite degree.

Launch

Time

45–60 minutes

Start by engaging the participants in a discussion of parabolas, quadratic functions, and lines of symmetry.

Ask participants to define each of these terms: *parabola, function, quadratic function,* and *line of symmetry.* Somewhere in the conversation may emerge a locus definition of a parabola as, in the plane, the set of points each of which is equidistant from a given point and a given line.

To illustrate the locus definition of *parabola*, demonstrate a dynamic geometry production of a parabola as shown in the directions (see section entitled Construction of Parabola Using the Locus Definition as well as Figures 4.1, 4.2, and 4.3). Have participants construct their own parabolas, using dynamic geometry as described subsequently in this section.

Construction of parabola using the locus definition

First, using a dynamic geometry environment, construct a line (called a *directrix*) and a point, F (called the *focus*), not on the directrix. Then construct a point D on the directrix and the segment DF. Finally construct the perpendicular bisector of segment DF. Call it *b*. The result is shown in Figure 4.1.

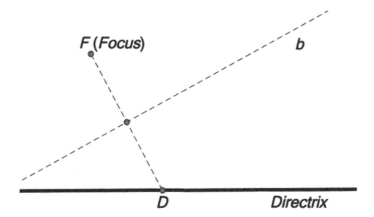

FIGURE 4.1.   Result of initial steps in the construction of a parabola using the locus definition.

Next, construct a perpendicular to the directrix through the point D. Identify the intersection of the perpendicular and *b* as the point P. The result of this construction is shown in Figure 4.2.

Finally, drag the point D along the directrix, tracing the point P as D is dragged. The resulting trace (shown by the thick curve in Figure 4.3) is a parabola with focus and directrix as labeled.

Points of discussion

1.   Ask participants to explain why this construction results in the set of points in a plane that are equidistant from a given point and a given line. The logic of the construction will likely not be immediately obvious to participants, but understanding the construction likely will enhance their understanding of the locus definition of *parabola*. Participants' explanations should address how for each given point, D, on the directrix, the fact that P is located on the perpendicular bisector of the segment with endpoints D and F ensures that the point P will be equidistant from the points F and D. Their explanations also should note that the fact that P is on a perpendicular to the directrix line ensures that it represents point P's distance from the directrix line.

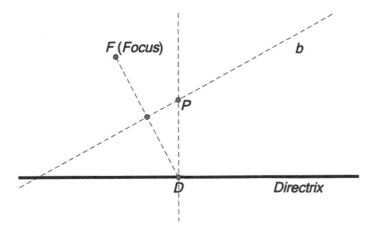

FIGURE 4.2.   Construction of a perpendicular to the directrix.

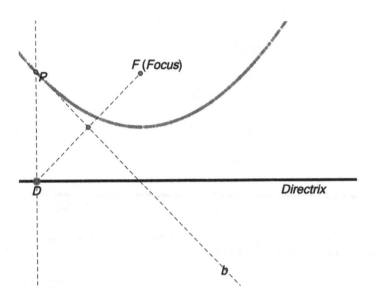

FIGURE 4.3.   Trace of point P as D is dragged along the directrix.

2.   Ask participants to determine the line of symmetry for their graph. Ask them to demonstrate that the line they have found is a good candidate for the line of symmetry. Although participants may address this question informally, it may be fruitful to encourage them to call up and use a definition of the line of symmetry for a vertical line, reinforcing the mathematical practice of reasoning from definitions.

3.   Ask participants to discuss any relationships they see between the line of symmetry and the focus as well as any relationships they see between the line of symmetry and the directrix. Questions about such relationships enhance participants' ability to make connections among related properties of a mathematical object.

4.   Ask participants to determine under what conditions a line of symmetry for a parabola is vertical, horizontal, slanted upward to the right, or slanted upward to the left. Questions such as these are likely to alert participants to the role of conditions in mathematical analysis.

In the Launch, establish that the Prompt refers to quadratic functions, rather than all parabolic figures. The distinction is that, in the standard, rectangular Cartesian coordinate system, quadratic functions graph as parabolas with a vertical axis of symmetry and a horizontal directrix.

Activity 1. Participants read the Prompt and make sense of the formula for the axis of symmetry of a parabola.

After participants read the Prompt, have groups talk about how they would make sense of the formula for the axis of symmetry given in the Prompt. In a whole-group setting, have participants present their explanations in an order of increasing complexity of thinking. For example, the explanations might be ordered from those that involve only early secondary school mathematics to those that involve calculus. Remind participants of the quadratic formula in the form $x = \dfrac{-b}{2a} \pm \dfrac{\sqrt{b^2 - 4ac}}{2a}$. Ask them to explain the meaning of the two addends and to discuss the role of the " $\pm$ " in the symmetry of the function graph.

Time

30 minutes

Anticipated participant responses

Participants are likely to share stories that are not directly related to finding the equation for the line of symmetry. For those stories that make a mathematical point and are related to graphing quadratic functions, record them on a newsprint "parking lot" to return to later if there is time.

Facilitating the activity

Read the Prompt to the participants and display an enlarged image of it on a screen or hand out a copy of the Prompt to the participants. Ask participants whether their students have asked similar questions when working with graphs of quadratic functions. Have them share these stories. Have groups talk about how they would respond to the Prompt. In a whole-group setting, have participants present their solutions.

Activity 2. In groups, participants discuss the assumptions that are made in each Focus.

Time

30 minutes

Anticipated participant responses

Participants may have difficulty identifying assumptions, and some of the ones they identify may actually be well supported in the arguments presented in the Foci. It will be important to provide opportunities for participants to vet their arguments and receive feedback on them from others in their groups.

Facilitating the activity

Have each group of participants read one of the Foci and identify an assumption. If they do not identify an assumption, offer them the one of the following that is connected to their assigned Focus.

- Focus 1 relies on the assumption that the graph of a quadratic function is symmetric about a line $y = k$, for some real number k.

- In Focus 2, the assumption is made that the critical value determines the equation for line of symmetry.

- In Focus 3, the assumption is made that neither a stretch or contraction (by a factor of a) nor a vertical translation through $\dfrac{b^2}{4a}$ units in the positive direction (or any vertical translation) affects the line of symmetry.

- Focus 4 uses Viète's formula to make the claim that the sum of the zeros of a quadratic function of the form $f(x) = ax^2 + bx + c$, with $a \neq 0$, is $-\dfrac{b}{a}$.

Ask each group to take on either the assumption they have located or one you have given them and ask them to provide corroborating evidence of its truth. Indicate to them that you are not necessarily requiring a proof but rather evidence that is consistent with the assumption.

Have groups share their results. Conduct a whole-group discussion on how the assumptions might be treated in a school mathematics setting. Participants may share ways in which they have had students encounter and investigate these assumptions. Some assumptions are the topics of Activities 4 and 5.

Activity 3. Participants read the Foci and, in groups, identify the mathematical background that the student teacher in the Prompt might draw upon or that a student might need for each of the Foci to be appropriate for him or her.

Time

30 minutes

Anticipated participant responses

Because of the application of derivatives in Focus 2, participants may see Focus 2 as appropriate only for students who have completed or are enrolled in a calculus course. They may see Focus 1 and Focus 3 as appropriate only for students at the precalculus level or beyond because of the intensive symbolic manipulation and the refined application of symmetry and interpretation of transformations. They may reserve Focus 4 for students engaged in mathematics competitions because Viète's formula is rarely taught in the high school curriculum.

Facilitating the activity

Regroup participants by courses taught (e.g., Algebra 1; Algebra 2; Geometry; Precalculus; Calculus; Integrated Mathematics 1, 2, 3, and 4). Ask each group to discuss how students in their designated course would approach one or more of the different solutions. Ask them to identify the needed understandings. Encourage them to say how they might challenge their students to consider more than one approach. Take a poll concerning what specific background students would need to be able to understand, and then to generate, each of the solutions. Ask participants to generate questions appropriate for students in their designated course about the relationships among the mathematical topics such as symmetry, trans-

formation, and calculus.

Key points about the Foci

- Focus 1. The argument in Focus 1 relies on using the symmetry of the curve about $x = k$ and equating the values of

    $f(x + 1)$ and $f(x - 1)$. The following argument is independent of a particular choice of directrix or line of symmetry.

    One can locate two distinct points with x-values $k + d$ and $k - d$ along the directrix, each of which is d units

    from k. One can also express the coordinates of P and Q, the corresponding points on the graph (see Figure 4.4).

    Symmetry about the line $x = k$ requires P and Q to have the same second coordinates. That is, $f(k + d) = f(k - d)$.

$$f(k+d) = f(k-d)$$
$$a(k+d)^2 + b(k+d) + c = a(k-d)^2 + b(k-d) + c$$
$$a(k+d)^2 + bd = a(k-d)^2 - bd$$
$$a(k^2 + 2kd + d^2) + bd = a(k^2 - 2kd + d^2) - bd$$
$$2adk + bd = -2akd - bd$$
$$4akd = -2bd$$
$$k = -\frac{b}{2a}$$

- Focus 2 uses the concept of derivative to derive the equation of the line of symmetry. The argument relies on the

    fact that the line of symmetry for a quadratic function is parallel to the y-axis and contains the vertex of the related

    parabola, where the vertex is the minimum or maximum point of the parabola.

- Key to Focus 3 is the recognition of which features of a function are preserved under which transformations. In

    this particular case, the line of symmetry is not affected by the given transformations. As the sequence of trans-

    formations suggested by $g(x) = v_1(f(h_2(x - h_1))) + v_2$ is performed, it is important to notice that h_1 acts on the input

    value x and h_2 acts on that result before the function f is applied. The parameters v_1 and v_2 act on results after the

    function f is applied.

- Focus 4 uses Viète's formulas for relationships between the coefficients of a quadratic polynomial and the sums

    and products of its roots. Teachers are likely not to be aware of the generalization of this formula to polynomials

    of degree n. Participants might investigate this formula as it applies to polynomials of smaller degree (degrees 2,

    3, and, perhaps, 4). The following, paraphrased from Weisstein (n.d.), gives a symbolic argument for the formula

    as it applies to polynomials of degree n as well as to the special cases of quadratic and cubic polynomials:

Let s_i be the sum of the products of distinct polynomial roots r_j of the polynomial equation of degree n.

$$a_n x^n + a_{n-1} x^{n-1} + \cdots + a_1 x + a_0 = 0$$

where the roots are taken i at a time. For example, the first few values of s_i are

$$s_1 = r_1 + r_2 + r_3 + r_4 + \cdots$$
$$s_2 = r_1 r_2 + r_1 r_3 + r_1 r_4 + r_2 r_3 + \cdots$$
$$s_3 = r_1 r_2 r_3 + r_1 r_2 r_4 + r_2 r_3 r_4 + \cdots$$

and so on. Then Viète's formula states that

$$s_i = (-1)^i \frac{a_{n-i}}{a_n}$$

The theorem was proved by Viète (also known as Vieta) for positive roots only, and the general theorem was proved by Girard.

The result can be seen in the context of a polynomial of degree two (a quadratic polynomial). This can be seen for a quadratic polynomial by multiplying out,

$$a_2 x^2 + a_1 x + a_0 = a_2(x - r_1)(x - r_2)$$
$$= a_2[x^2 - (r_1 + r_2)x + r_1 r_2],$$

so

$$s_1 = \sum_{i=1}^{2} r_i$$
$$= r_1 + r_2$$
$$= -\frac{a_1}{a_2}$$

$$s_2 = \sum_{\substack{i,j=1 \\ i \neq j}}^{2} r_i r_j$$
$$= r_1 r_2$$
$$= \frac{a_0}{a_2}$$

The following shows how Viète's formulas relate to what students learn about factoring: The sum of the roots of a quadratic function $y = ax^2 + bx + c$ is $-\frac{b}{a}$ and the product of the roots is $\frac{c}{a}$.

$$ax^2 + bx + c = a \cdot (x - r) \cdot (x - s)$$
$$= a \cdot (x^2 - (r + s) \cdot x + r \cdot s)$$

$$s_1 = \sum_{i=1}^{2} r_i$$
$$= r + s$$
$$= -\frac{b}{a}$$

$$s_2 = \sum_{\substack{i,j=1 \\ i \neq j}}^{2} r_i r_j$$
$$= r \cdot s$$
$$= \frac{c}{a}$$

Similarly, for a third-degree polynomial,

$$a_3 x^3 + a_2 x^2 + a_1 x + a_0 = a_3 (x - r_1)(x - r_2)(x - r_3)$$
$$= a_3 [x^3 - (r_1 + r_2 + r_3)x^2 + (r_1 r_2 + r_1 r_3 + r_2 r_3)x - r_1 r_2 r_3],$$

so

$$s_1 = \sum_{i=1}^{3} r_i = -\frac{a_2}{a_3}$$
$$s_2 = r_1 + r_2$$
$$= -\frac{a_1}{a_2}$$

$$s_2 = \sum_{\substack{i,j=1 \\ i \neq j}}^{3} r_i r_j$$
$$= r_1 r_2 + r_1 r_3 + r_2 r_3$$
$$= \frac{a_1}{a_3}$$

$$s_3 = \sum_{\substack{i,j,k=1 \\ i<j<k}}^{3} r_i r_j r_k$$
$$= r_1 r_2 r_3$$
$$= -\frac{a_0}{a_3}$$

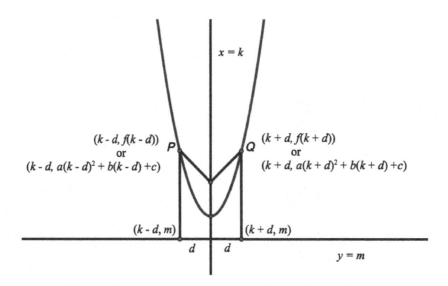

FIGURE 4.4. Location of two distinct points with x-values of $k + d$ and $k - d$.

Key points for discussion

- The mathematical knowledge needed to carry out each of the justifications for the equation for the line of symmetry differs. Ask participants to discuss similarities and differences among the justifications.

- The mathematical development in this Situation involved parabolas with vertical lines of symmetry. Parabolas with horizontal lines of symmetry can be developed drawing on similar mathematical ideas, whereas parabolas with slanted lines of symmetry involve a composition of transformations. Ask participants how they might develop the equation for a line of symmetry if the parabola had a horizontal line of symmetry or it the parabola had a slanted line of symmetry.

- This Facilitator's Guide suggests asking participants to examine the effects of transformations on graphs that have vertical lines of symmetry. An investigation of the effects of transformations on horizontal lines of symmetry is likely to involve similar mathematical ideas, whereas an algebraic analysis of the effects of transformations on slanted lines of symmetry are more complicated and are likely to challenge even mathematically savvy teachers. If appropriate to the background of the participants, ask them to discuss the effects of transformations such as stretches, contractions, and translations for figures that have a horizontal or slanted axis of symmetry.

- Participants are likely to have taught their students something about lines of symmetry for quadratic functions. Ask participants how they have developed or might develop an understanding of graphing quadratics over the course of a high school mathematics program.

- Students are likely to need to call on different mathematics to engage in each of the Foci. Ask participants to discuss what evidence of students' mathematical thinking would be needed to support each of the four Foci.

- Using the Foci in this Situation, teachers can develop lessons that engage students in different mathematical practices. Ask participants to identify ways to use mathematical ideas from this Situation to engage students in different mathematical practices.

Activity 4. Groups of participants investigate how each type of transformation in Focus 3 (horizontal or vertical translation, horizontal or vertical stretch, horizontal or vertical contraction) affects the graph of the function.

Time

30 minutes

Anticipated participant responses

Participants may need to work through a specific example in order to understand the effects of transformations on properties of the graph of a function. Some of the effects are more easily understood than others. Participants are likely to understand readily why the graph of $f(x) + k$ (for $k > 0$) is a vertical translation upward of the graph of $f(x)$. It would not be unexpected for participants to have some difficulty with why a horizontal translation of a function works the way that it does (e.g., $f(x + 2)$ represents a horizontal translation of $f(x)$ two units in the negative direction rather than a horizontal translation of $f(x)$ two units in the positive direction).

Facilitating the activity

Ask participants to use a specific quadratic function and their graphing utility to corroborate the claims made about the effects of different transformations on the line of symmetry. Ask them to identify difficulties students might have with this conclusion and how they might work with students to generate an understanding of Focus 3. As an extension, you might ask them to identify other properties of a parabola and determine how those properties would be affected by translations, stretches, and contractions.

Activity 5. As time allows, groups of participants can consider how the content of the Foci might be used to connect with other topics in the secondary school curriculum. For example, how do Viète's formulas for the roots of a polynomial relate to techniques that students use to factor quadratic expressions?

Time

1.5 hours

Anticipated participant responses

Participants may need an introduction to the CAS technology such as the TI-Nspire software or symbolic calculator. If so, lead them through how to enter polynomial expressions and how to use the Expand command, and how to show a result that extends more than a screen width. Provide "help sheets" with commands, as needed.

Facilitating the activity

Relationship among coefficients of a quadratic expression, the roots of the associated quadratic function, and Viète's formulas.

Participants may be familiar with the relationship between the coefficients of a quadratic expression and the roots of the associated quadratic function, but they may not realize it as an application of Viète's formulas. That is, they may know that a way to show that if $r + s = -\dfrac{b}{a} r + s = -\dfrac{b}{a}$ and $r \times s = \dfrac{c}{a}$, then r and s are solutions to $ax^2 + bx + c = 0$. Here is an argument to support that claim:

$$\text{Suppose } r + s = -\frac{b}{a} \text{ and } r \cdot s = \frac{c}{a}.$$
$$\text{Then } ax^2 + bx + c = 0, a \neq 0$$
$$\Rightarrow ax^2 - a\left(-\frac{b}{a}\right)x + a \cdot \left(\frac{c}{a}\right) = 0, a \neq 0$$
$$\Rightarrow ax^2 - a(r+s)x + a \cdot r \cdot s = 0, a \neq 0$$
$$\Rightarrow a \cdot (x-r) \cdot (x-s).$$
$$\text{So } r \text{ and } s \text{ are solutions to } ax^2 + bx + c = 0, a \neq 0.$$

This result can provide participants an entry into generalizing the result to polynomials of higher degree. The following development can be used to give participants significant experience in one of the Common Core Mathematical Practices, Look for and Make Use of Structure, at a level that they are likely to find accessible but challenging. The extension of this activity to work with polynomials with literal coefficients can engage the participants in another Common Core Mathematical Practice, Reason Abstractly and Quantitatively.

Relationship among coefficients of a polynomial of degree n, the roots of the associated polynomial function, and Viète's formulas.

The goal of the following activity is to develop an understanding of the relationship between the coefficients of the terms of a polynomial and its zeros. The activity starts with a specific polynomial and then extends to the general case.

Start by having participants use a CAS to expand the following products:

$$(x - 1)(x - 2)$$

$$(x - 1)(x - 2)(x - 3)$$

$$(x - 1)(x - 2)(x - 3)(x - 4)$$

$$(x - 1)(x - 2)(x - 3)(x - 4)(x - 5).$$

They should obtain results such as

$$(x - 1)(x - 2) = x^2 - 3x + 2$$

$$(x - 1)(x - 2)(x - 3) = x^3 - 6x^2 + 11x - 6$$

$$(x - 1)(x - 2)(x - 3)(x - 4) = x^4 - 10x^3 + 35x^2 - 50x + 24$$

$$(x - 1)(x - 2)(x - 3)(x - 4)(x - 5) = x^5 - 15x^4 + 85x^3 - 225x^2 + 274x - 120.$$

Ask them to predict (without using the CAS) the expanded product of

$$(x - 1)(x - 2)(x - 3)(x - 4)(x - 5)(x - 6).$$

If, after a few minutes, they have not generated

$$x^6 - 21x^5 + 175x^4 - 735x^3 + 1624x^2 - 1764x + 720,$$

focus their attention on the sequence of the absolute values of the coefficients of the second term in each expansion (3, 6, 10, 15) and the absolute values of the last term in each expansion (2, 6, 24, 120).

Participants may recognize (3, 6, 10, 15) as an ordered 4-tuple of triangular numbers and, hence, as the sums of the first 2, 3, 4, and 5 natural numbers, respectively. They may also recognize (2, 6, 24, 120) as a sequence consisting of the second through fifth factorials. If the discussion evolves this way, ask participants to discuss in their groups what accounts for these particular patterns of results.

Ask them to discuss in their groups any other patterns they see. If some groups (or individuals) have discovered the patterns, ask them not to share it with the whole group but to pose a question for the whole group that will not give it away but will help others in the group to think about the patterns.

Participants who have generated the sequence of specific products may gain additional insight by generating the following sequence of products, using a CAS,[6] thus engaging in reasoning abstractly:

$$(x - a)(x - b)$$

$$(x - a)(x - b)(x - c)$$

$$(x - a)(x - b)(x - c)(x - d).$$

The following are equivalent to the answers they should acquire:

$$(x - a)(x - b) = x^2 - ax - bx + ab = x^2 - (a + b)x + ab$$

$$(x - a)(x - b)(x - c) = x^3 - ax^2 - bx^2 - cx^2 + abx + acx + bcx - abc = x^3 - (a + b + c)x^2 + (ab + ac + bc)x - abc$$

$$(x - a)(x - b)(x - c)(x - d) = x^4 - ax^3 - bx^3 - cx^3 - dx^3$$

$$+ abx^2 + acx^2 + adx^2 + bcx^2 + bdx^2 + cdx^2$$

$$- abcx - abdx - acdx - bcdx + abcd$$

$$= x^4 - (a + b + c + d)x^3 + (ab + ac + ad + bc + bd + cd)x^2 - (abc + abd + acd + bcd)x + abcd.$$

Ask them to predict the expanded product of $(x - a)(x - b)(x - c)(x - d)(x - e)$, which should be equivalent to:

$$x^5 - (a + b + c + d + e)x^4 + (ab + ac + ad + ae + bc + bd + be + cd + ce + de)x^3$$

$$- (abc + abd + abe + acd + ace + ade + bcd + bce + bde + cde)x^2$$

$$+ (abcd + abce + abde + acde + bcde)x$$

$$- abcde.$$

Ask groups to generate a way to describe the coefficients. Although participants may have difficulty articulating the general relationships, they can be encouraged to look at the quadratic and cubic cases. For the quadratic polynomial that is the product of two of these binomials, the coefficient of x is the opposite of the sum of the roots, and the constant is the product of the roots. For the cubic polynomial that is the product of three of these binomials, the coefficient of x^2 is the opposite of the sum of the roots, the coefficient of *x* is the sum of the products of all possible pairs of roots, and the constant is the opposite of the product of all the roots. Once participants have articulated the pattern for the coefficients in the quadratic and cubic cases, they may find it easier to generalize the pattern to the quartic case and recognize the coefficient of the x^3 term as the opposite of the sum of all possible three-way products of roots.

Few participants likely can articulate the case of the product of *n* such binomials. More advanced participants, however, might be able to interpret the following articulation of the relationship. The coefficient of x^p in the product of *n* binomials of the form $(x - k)$ is the sum of the products of the *k*'s taken $(n - p)$ at a time (e.g., for *n* = 5, the coefficient of x^3 in $(x - a)(x - b)(x - c)(x - d)(x - e)$ is the sum of the products of *a*, *b*, *c*, *d*, and *e*, taken $5 - 3$, or 2, at a time).

The explanations in this professional learning document are intended for high school teachers rather than for their students. This exploration in its fullest form is likely to be beyond the reach of all but the most advanced high school students. However, a careful development of the case of a cubic or quartic polynomial can develop students' understanding of where the coefficients come from. Ask groups to generate a way to explain to advanced high school algebra students why these patterns exist in a cubic or quartic polynomial. An interesting extension can center on students reasoning about the number of addends for each coefficient. That is, teachers can focus students' attention on the number of products that are summed to determine each coefficient (e.g., In the expansion of $(x - a)(x - b)(x - c)(x - d)$, why is the coefficient of x^2 the sum of six terms?).

This discussion of the number of products that are summed to determine the coefficient of various powers of *x* leads to an extension of patterns to polynomials of higher degree. Participants may generalize their observations to the expansion of $(x - a)(x - b)(x - c)(x - d)(x - e)$, and the observation that, for example, 10 terms are added to arrive at the coefficient of x^2. Or, they may develop a rationale for such observations as the fact that the terms in these expansions alternate in sign.

Reflect and assess learning

Time

30–60 minutes

Ask participants what background they might develop in students before engaging them in the construction of a parabola using the locus definition. Depending on the size of the group, divide the group into "course-alike" subgroups for this discussion.

- Ask participants to reflect on their experience with Viète's formulas and determine what, if any, engagement with the formulas might be appropriate for the students they teach at various levels.

- Ask participants who teach the same level (e.g., Algebra 1, Algebra 2, Geometry, Precalculus, Calculus) to discuss whether some part of this exploration might be used with their students. Participants who deem such an exploration as appropriate for their students could produce a poster that illustrates how they would approach this prompt in their own classrooms. Arrange for a gallery walk during which participants could make suggestions about this approach (by attaching comments on sticky notes).

- Ask participants to self-assess their current understanding of the four Foci prior to discussion of the Situation and then reflect on their growth after the discussion.

Reflection questions

1. Has our work with this Situation caused you to consider or reconsider any aspects of your own thinking and/or practice about graphing quadratic functions? Explain.

2. Has our work with this Situation caused you to reconsider any aspects of your students' mathematical learning about graphing quadratic functions? Explain.

3. Has our work with this Situation engaged us in mathematical practices? Explain.

4. What additional questions has our work with this Situation raised for you?

5. What mathematical points were new to you? What mathematical ideas in this Situation would you like to pursue either on your own or with students?

RESOURCES

Websites

For a video illustrating the construction of a parabola using the locus definition, see www.youtube.com/watch?v=xUQF5ZqbmeU
For a discussion of the construction of a parabola using the locus definition, see jwilson.coe.uga.edu/emt725/Class/McAdams/MA%20
668%20WRITE%20UPS/ASSIGNMENT%206/Assignment%20Six.html
The Common Core State Standards for Mathematics can be accessed at http://www.corestandards.org/Math

NOTES

1. The Graphing Quadratic Functions Situation is one of the Situations presented in *Mathematical Understanding for Secondary Teaching: A Framework and Classroom-Based Situations* (Heid & Wilson, 2015).

2. This Situation appears on pp. 257–261 of Heid and Wilson (2015). It is reprinted with permission.

3. For each standard, we have suggested questions (those in italicized, boldface type) that might serve as follow-up questions to address the mathematics of the standard.

4. In the Common Core State Standards for Mathematics, additional mathematics that students should learn in order to take advanced courses such as calculus, advanced statistics, or discrete mathematics is indicated by (+).

5. See the Appendix for the list of the Standards for Mathematical Practice in the Common Core State Standards for Mathematics (National Governors Association Center for Best Practices and Council of Chief State School Officers, 2010).

6. Participants may have to be reminded to clear values for the variables before they start their exploration.

REFERENCES

Heid, M. K. & Wilson, P. W. (with G. W. Blume). (Eds.). (2015). *Mathematical understanding for secondary teaching: A framework and classroom-based situations*. Charlotte, NC: Information Age.

National Governors Association Center for Best Practices & Council of Chief State School Officers. (2010). *Common core state standards for mathematics*. Washington, DC: Authors.

Weisstein, E. W. (n.d.). *Vieta's formulas*. Retrieved from www.mathworld.Wolfram.com/VietasFormulas.html.

CHAPTER 5

FACILITATOR'S GUIDE
FOR
CIRCUMSCRIBING POLYGONS

Situation 34 From the MACMTL–CPTM Situations Project[1]

Patricia S. Wilson and M. Suzanne Mitchell

Classes of polygons are a common topic in a secondary school mathematics curriculum. Students are expected to know that polygons are planar figures and how different classes of polygons are related to each other. For example, students learn that squares are rhombi with congruent angles, and rhombi are parallelograms with four congruent sides. Polygons can also be sorted into classes based on criteria other than the number and measure of angles and sides. In this Situation, participants identify and compare cyclic polygons, triangles, regular polygons, quadrilaterals, and rectangles. They also interpret and create statements regarding whether one class is a subset of another. This type of question leads to answers expressed as conditionals (e.g., If ABCD is a square, then ABCD is a rhombus.) and biconditionals (e.g., A figure is a regular quadrilateral if and only if the figure is a square.). The Situation encourages teachers to identify the difference between conditional statements and biconditional statements and differences in how such statements are proven. The setting for the Prompt in this Situation was a high

Facilitator's Guidebook for Use of Mathematics Situations in Professional Learning,
pages 85–114.

school Geometry class, but the content is relevant to geometry topics in Grades 6–12. Understanding circumscribing a polygon helps students visualize a circle as a locus of points in a plane equidistant from a common point. It helps students understand properties of polygons, including similarities and differences among different types of polygons.

OVERVIEW

Situation	Relevance
In a geometry class, after a discussion about circumscribing circles about triangles, a student asked, "Can you circumscribe a circle about any polygon?"	• Exploring polygons and their characteristics is common middle school and high school mathematics activity. • Understanding the nature of the circumcenter helps students circumscribe polygons, know properties of circumscribed polygons, and appreciate the properties of the perpendicular bisector of a line segment. • Using deductive reasoning to build a bridge from circumscribing triangles to circumscribing specific types of polygons is a way to integrate learning about polygons and proving hypotheses about polygons.
Goals	Key Mathematical Topics
• Increase teachers' understanding of characteristics of polygons. • Increase teachers' understanding of circumscribing polygons. • Increase teachers' understanding of biconditional proofs and deductive reasoning in general.	• A polygon that can be circumscribed by a circle is a cyclic polygon. • Every triangle is cyclic. This fact is key to establishing conditions for deciding which polygons are cyclic. • A convex quadrilateral in a plane is cyclic if and only if its opposite angles are supplementary. • Every planar regular polygon is cyclic. • Concave polygons in a plane are not cyclic. • A statement is a biconditional statement if it is the equivalent of the combination of a conditional statement and the converse of that statement. • To prove a biconditional requires proving a conditional statement and its converse.

The Situation under consideration in this chapter follows (highlighted with a gray background).[2]

CIRCUMSCRIBING POLYGONS

Situation 34 From the MACMTL-CPTM Situations Project

Shari Reed, AnnaMarie Conner, Heather Johnson,
M. Kathleen Heid, Bob Allen, Shiv Karunakaran,
Sarah Donaldson, and Brian Gleason

PROMPT

In a geometry class, after a discussion about circumscribing circles about triangles, a student asked, "Can you circumscribe a circle about any polygon?"

COMMENTARY

A polygon that can be circumscribed by a circle is called a *cyclic polygon.*[1] Not every polygon is cyclic, but there are infinitely many different cyclic polygons. This can be understood by considering a given circle and all the possibilities of how many points can be placed on the circle, and then connected to form a polygon. However, there are certain classes of polygons that are noteworthy because they are always cyclic. The conditions under which a circle circumscribes a given polygon are dependent upon the relationships among the angles, the sides, and the perpendicular bisectors of the sides of the polygon. The following Foci describe classes of cyclic polygons in order of the number of their sides: triangles, certain quadrilaterals, and regular polygons. Focus 3 provides one way to check whether a given polygon is cyclic: A polygon is cyclic if and only if the perpendicular bisectors of all its sides are concurrent. Although the inclusion of various geometries would provide interesting discussion, the Foci in this Situation are limited to Euclidean geometry in a plane.

MATHEMATICAL FOCI

Mathematical Focus 1

Every triangle is cyclic. This generalization is central to establishing a condition for other polygons to be cyclic.

Because the center of a circle is equidistant from all points on the circle (this distance is the radius), an *inscribed triangle* is one in which the three vertices of the triangle lie on the circumscribed circle. Conversely, the circle circumscribed about a particular triangle must have as its center the point that is equidistant from the vertices of the triangle. Circumscribing a circle about a triangle, then, requires finding a point that is equidistant from the three vertices of the triangle. This point is called the *circumcenter* of the triangle.

A point in a plane is equidistant from points A and B in that plane if and only if it lies on $\overline{AB}$'s perpendicular bisector that lies in that plane. A proof of this theorem is included in the Postcommentary. Because of this, consider perpendicular bisectors of the sides of a triangle to find the circumcenter. Given $\triangle ABC$ (see Figure 40.1), the perpendicular bisectors (in the plane of $\triangle ABC$) of segments AB and BC intersect the sides at D and E, respectively. $\overline{AB}$ and $\overline{BC}$ are not parallel, so lines that are perpendicular to them are not parallel. Therefore, the perpendicular bisectors of $\overline{AB}$ and $\overline{BC}$ must intersect at some point, call it P. P is equidistant from A and B because it lies on the perpendicular bisector of $\overline{AB}$, and P is equidistant from B and C because it lies on the perpendicular bisector of $\overline{BC}$. So P is equidistant from A, B, and C. That is, P is the circumcenter of $\triangle ABC$. So, given any set of three noncollinear points, it is possible to find the circumcenter of the triangle defined by those three points.

Another way to think about triangles being cyclic is to consider the equation of a circle in the coordinate plane: $(x - h)^2 + (y - k)^2 = r^2$ where (h, k) is the center of the circle, r is its radius, and every ordered pair (x, y) that satisfies the equation lies on the circle. To find a particular circle (that is, to find the three unknowns h, k, and r), one would need three equations. That is, if one had three ordered pairs (x, y) (i.e., three noncollinear points), one could determine the circle. This

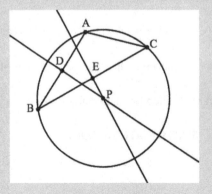

FIGURE 40.1. Location of the circumcenter, P, for triangle ABC.

is another way of showing that three noncollinear points determine a unique circle. Given those three points, one could find the circumcenter of the triangle defined by those points.

Mathematical Focus 2

A convex quadrilateral in a plane is cyclic if and only if its opposite angles are supplementary.

A convex quadrilateral is a quadrilateral in a plane such that no two points in the interior of the quadrilateral can be connected by a segment that intersects one of the sides.[2] Proving that two conditions (a quadrilateral being cyclic and its opposite angles being supplementary) are equivalent requires proving an implication and its converse. That is, to prove $A \Leftrightarrow B$ one must prove that $A \Rightarrow B$ and prove that $B \Rightarrow A$. This proof uses a logically equivalent construction of proving $A \Rightarrow B$ and then proving *not A $\Rightarrow$ not B*.

a.   First, prove that given a convex, cyclic quadrilateral, its opposite angles are supplementary. In the cyclic quadrilateral ABCD in Figure 40.2, $\angle ABC$ is opposite $\angle CDA$. Because the measure of an inscribed angle is half the measure of the arc in which it is inscribed,

$$m\angle ABC = (1/2)\, m(\text{arc}CDA) \text{ and } m\angle CDA = (1/2)\, m(\text{arc}ABC)\,^3$$

$$m\angle ABC + m\angle CDA = (1/2)\, m(\text{arc}CDA) + (1/2)\, m(\text{arc}ABC)$$

$$= (1/2)\, [m(\text{arc}CDA) + m(\text{arc}ABC)].$$

Because the union of arcs CDA and ABC is a circle, $m(\text{arc}CDA) + m(\text{arc}ABC) = 360°$. By substitution, $m\angle ABC + m\angle CDA = (1/2)(360) = 180°$. Therefore $\angle ABC$ and $\angle CDA$ are supplementary. (Angles BAD and DCB could be handled similarly.)

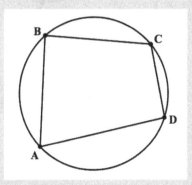

FIGURE 40.2.   Cyclic quadrilateral ABCD.

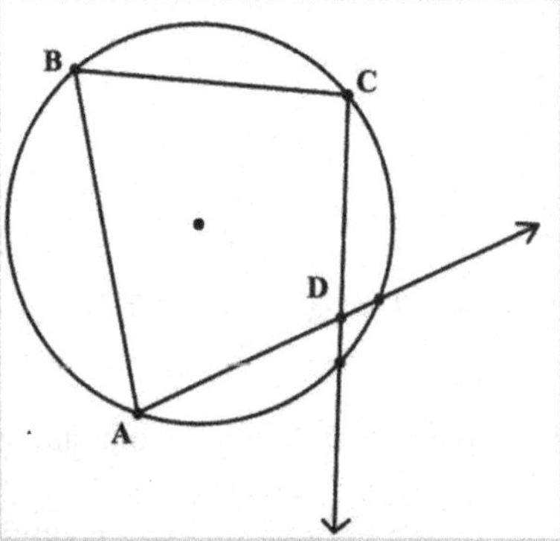

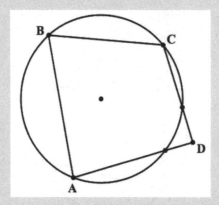

FIGURE 40.3. Convex quadrilateral ABCD; the circle defined by points A, B, and C; and point D in the interior of that circle.

b.   Next prove that if the opposite angles of a quadrilateral are supplementary, then the quadrilateral is cyclic. Begin

with convex quadrilateral ABCD such that angles BAD and DCB are supplementary and angles ABC and CDA are

supplementary. Draw the circle defined by points A, B, and C. (This circle can be constructed because three points

determine a circle—see Focus 1). Suppose D is located in the interior of the circle (see Figure 40.3). Then extend

segments AD and CD until they each intersect the circle. By the inscribed angle theorem, the sum of angles BAD

and DCB is less than 180 degrees (because together they subtend less than a whole circle[4]). This is a contradiction,

therefore D cannot be inside the circle.

Suppose instead that D is outside the circle (see Figure 40.4). In this case, the sum of the measures of angles BAD and

DCB will be greater than 180° because together they subtend more than a whole circle (they subtend more than the whole

FIGURE 40.4. Convex quadrilateral ABCD; the circle defined by points A, B, and C; and point D in the exterior of that circle.

circle, accounting for a portion of the circle twice). This is a contradiction, so D cannot be outside the circle. Because D cannot be inside or outside the circle, D must be on the circle. So, quadrilateral ABCD is cyclic.[5]

A corollary that is implied by the preceding result is that every rectangle is cyclic. Also, no parallelograms other than rectangles are cyclic. Moreover, every isosceles trapezoid is cyclic. [Note: A commonly used definition of *trapezoid* is that it is a quadrilateral with exactly one pair of parallel sides. However, trapezoids are defined by some sources as a quadrilateral with at least one set of parallel sides. If trapezoids are defined as in the latter definition, then every rectangle is an isosceles trapezoid.]

Mathematical Focus 3

There are four-sided figures in the plane that behave differently from convex quadrilaterals. Concave quadrilaterals are never cyclic, and a four-sided figure with nonsequential vertices is cyclic if and only if its "opposite" angles are congruent.

Cyclic quadrilaterals have opposite angles that are supplementary (see Focus 2). Consider a concave polygon such as quadrilateral EFGH in Figure 40.5. To show that quadrilateral EFGH is noncyclic, construct the circle passing through the points EFG (in general, through the 3 points that occur at the polygon's interior angles with measures that are not greater than 180°). Constructing this circle is possible because a concave quadrilateral will have exactly one interior angle greater than 180 degrees; in this case, it is angle EHG. Suppose that point H lies on this circle. But then the measure of (interior) angle EHG is less than 180 degrees because it cannot subtend the whole circle (much less even more than the whole circle), producing the contradiction that m∠EHG > 180° and m∠EHG < 180°. So, point H cannot lie on the circle passing through E, F, and G, so concave quadrilateral EFGH is noncyclic.

A quadrilateral is commonly defined as a polygon with four sides. Therefore it is important that the definition of polygon be clear. If the definition of polygon requires that it be a simple, closed figure (as it is in many high school mathematics

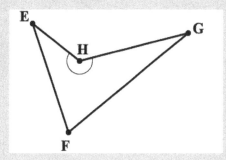

FIGURE 40.5. Concave quadrilateral EFGH.

textbooks), then the figures in the following discussion are not polygons, and therefore not quadrilaterals. However, if the definition requires only that a quadrilateral be a closed figure in a plane with four straight sides, then a quadrilateral with nonsequential vertices can be considered in this Situation.

A quadrilateral with nonsequential vertices is cyclic if and only if its "opposite" angles are congruent. Here, the vertices of a quadrilateral ABCD are called *nonsequential* if side AB intersects side CD. Also, the term *opposite angles* of a quadrilateral is meant to convey two interior angles of the quadrilateral that do not share a common side. Such a quadrilateral has sides that "cross," as seen in Figure 40.6. "Opposite" angles in Figure 40.6, for example, are angles NMP and PON or angles ONM and MPO.

In this case, it must first be proved that if a quadrilateral with nonsequential vertices is cyclic, then its "opposite" angles are congruent. Consider quadrilateral MNOP in Figure 40.6. Because angles NMP and PON lie on the circle and intercept the same arc (arc NP), they are congruent. In the same way, angles ONM and MPO both intercept arc MO, so they are congruent. Note that the sum of the measures of the interior angles of this type of quadrilateral is not 360°.

Next, prove that if "opposite" angles of a quadrilateral with nonsequential vertices are congruent, then the quadrilateral is cyclic. The same strategy that was used for the converse in Part a of Mathematical Focus 2 can be used here: Begin with quadrilateral MNOP such that angles NMP and PON are congruent, angles ONM and MPO are congruent, and all the vertices except P lie on circle r (Figure 40.7). If P is inside circle r, then extend sides MP and OP to intersect circle r. Then angles NMP and PON subtend different size arcs, so their measures are not equal, which contradicts the statement that angles NMP and PON are congruent. If P is outside circle r, then side MP intersects circle r in a different point than does side OP, so angles NMP and PON are again noncongruent, which again is a contradiction. Therefore, P must lie on circle r.

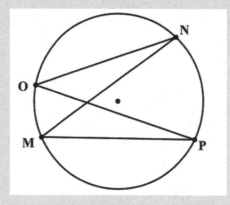

FIGURE 40.6.   Quadrilateral MNOP with nonsequential vertices.

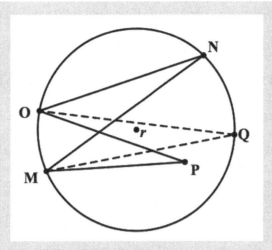

FIGURE 40.7. Quadrilateral MNOP with all vertices except P on circle *r*.

Mathematical Focus 4

Every planar regular polygon is cyclic. However, not every cyclic polygon is regular.

As was discussed in Focus 1, a point is equidistant from two points, A and B, if and only if it lies on the perpendicular bisector of the segment whose endpoints are A and B. For a polygon to be cyclic, there must be a circle that passes through all of its vertices. In other words, there must be a single point that is equidistant from all the vertices. This point must lie on the perpendicular bisectors of all the sides of the polygon. For a point to lie on all these bisectors, the bisectors must be concurrent, and the point of concurrency will be the circumcenter of the polygon. Because the statement about equidistance and the perpendicular bisector is biconditional, a biconditional statement can be made about the concurrency of the perpendicular bisectors of the sides of a polygon. That is, the perpendicular bisectors of the sides of a polygon are concurrent if and only if the polygon is cyclic.

Every triangle is a cyclic polygon, as was established in Focus 2. The question remains as to which other polygons are cyclic. By examining the perpendicular bisectors of the sides of a polygon, one can determine a set of conditions on a polygon that is sufficient to conclude whether a circle can circumscribe that polygon. In particular, one can show that every regular polygon is cyclic.

Having established that if the perpendicular bisectors of the sides of a polygon are concurrent, the polygon is cyclic, it remains to show that the perpendicular bisectors of the sides of a regular polygon are concurrent and conclude that every regular polygon is cyclic.

By definition, a *regular polygon* is an equilateral and equiangular *n*-sided polygon. Consider a regular polygon with adjacent vertices, A, B, C, and D. Let P be the point of intersection of the perpendicular bisectors ($\overline{FP}$ and $\overline{GP}$, respectively) of $\overline{AB}$ and $\overline{BC}$. It can be shown that $\triangle AFP \cong \triangle BFP \cong \triangle BGP \cong \triangle CGP$ using the fact that P is equidistant from A, B, and C, and using the HL (hypotenuse–leg) congruence theorem. Construct $\overline{PH}$ perpendicular to $\overline{DC}$ and consider $\triangle HCP$. $\angle FBG \cong \angle GCH$ because the polygon is equiangular. $\angle FBP \cong \angle GCP$ because $\triangle BFP \cong \triangle CGP$. By angle subtraction, $\angle GBP \cong \angle HCP$, so $\triangle FBP \cong \triangle HCP$ by AAS (angle–angle–side triangle congruence). Congruent triangles establish that HC = FB, and because the polygon is equilateral, AB = CD. Also, FB = (½)AB because F is the midpoint of AB. So HC = (½)CD by substitution. Because C, H, and D are collinear, HC = HD. Thus, $\overline{PH}$ is the perpendicular bisector of $\overline{DC}$. The argument can be extended to successive vertices of the polygon, resulting in establishing that each of the perpendicular bisectors of the sides contains the point P (Figure 40.8). That is, the perpendicular bisectors are concurrent. Note that this argument does break down if the polygon in question is not regular. If the polygon is not regular (specifically if the sides are not equal to each other) one cannot prove that all the triangles listed previously are congruent.

Therefore, every regular polygon has concurrent perpendicular bisectors and therefore is cyclic. Although every regular polygon is cyclic, it is not true that every cyclic polygon is regular. For example, every triangle is cyclic, but every triangle is not an equilateral triangle and thus not regular. So it is possible to have polygons that are not regular but are cyclic (see Figure 40.9).

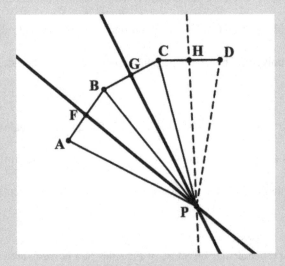

FIGURE 40.8.   Regular polygon ABCD··· with point P the intersection of the perpendicular bisectors of its sides.

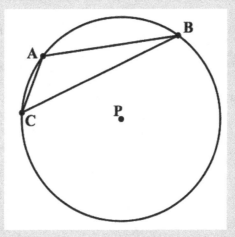

FIGURE 40.9.   A cyclic polygon, ΔABC, that
is not regular.

POSTCOMMENTARY

In making mathematical statements, it is important to recognize which are biconditional and which are not. In Focus 2, the proof of a statement and its converse constituted a proof of the biconditional. In Focus 4, the converse of the statement proved is not true, and so the biconditional is not true. In Focus 1, the use of the biconditional was not as overt. That Focus drew upon the property that each point on a perpendicular bisector of a segment is equidistant from the endpoints of the segment. Later in the Focus, establishing uniqueness was based upon the result that, in a plane, points that are equidistant from the endpoints of a segment lie on the perpendicular bisector of the segment. Because this statement was the converse of a previously established property, it requires a proof. That proof follows here:

A line is a perpendicular bisector of a segment AB if and only if it is the set of all points in a plane containing segment AB that are equidistant from A and B.

a.   If a point lies on the perpendicular bisector of a line segment AB, then it is equidistant from A and B.

Let M be the midpoint of segment AB, and let P be a point that lies on the perpendicular bisector of segment AB such that P does not lie on segment AB (see Figure 40.10). By the definition of perpendicular bisector, AM = BM and m∠BMP = m∠AMP = 90°. Now consider ΔAMP and ΔBMP. These triangles share side PM, so they are congruent by SAS (side–angle–side triangle congruence). Therefore PA = PB.

b.   If a point is equidistant from A and B then it lies on the perpendicular bisector of the line segment AB.

Let P be a point not on segment AB such that PA = PB. This means ΔPAB is isosceles and its base angles are congruent, so m∠BAP = m∠PBA. Let M be the midpoint of segment AB. Because triangles PAM and PBM share side PM, they are congruent by SAS. Because ∠BMA is a straight angle, m∠PMA + m∠BMP = 180°. Also, m∠PMA = m∠BMP because ΔPAM ≅ ΔPBM. Thus,

$$m\angle PMA + m\angle BMP = 180°$$

$$\Rightarrow 2(m\angle PMA) = 180°$$

$$\Rightarrow m\angle PMA = 90°$$

Therefore segment PM is a perpendicular bisector of segment AB.

NOTES

1. See Leung and Lopez-Real (2002) for additional discussion of conditions under which a polygon may or may not be cyclic.

2. This definition can be generalized to that of a convex object: An object O in a vector space is said to be *convex* if for any two points p and q in O, the line segment from p to q lies entirely inside O (meaning in its interior and/or on its boundary). The line segment can be described as the set of points $(1 - t) \cdot p + t \cdot q$, for t in [0, 1].

3. Arc measures and angle measures are assumed to be in degrees.

4. The claim that these angles subtend less than the whole circle relies on the figure, which shows that part of the circle is unaccounted for in the two subtended arcs.

5. One can give a nonpicture-based proof of the fact that if D is exterior to the circle, then the corresponding subtended arcs can be shown to have an arc of the circle as intersection, and cover more than the entire circle. And similarly, if D is interior to the circle, then the union of the corresponding subtended arcs can be shown to omit an arc of the circle.

REFERENCE

Leung, A., & Lopez-Real, G. (2002). Theorem justification and acquisition in dynamic geometry: A case of proof by contradiction. *International Journal of Computers for Mathematical Learning, 7,* 145–165.

CONNECTION TO STANDARDS

Learning experiences in mathematics for teachers are well positioned when teachers are aware of connections to the standards for which their students are accountable. These standards differ across states and provinces as well as across countries, although there are commonalities across different sets of standards. Although this document cannot possibly address all the existing sets of standards across states, provinces, and countries, an example follows that illustrates how the proposed professional learning might address one particular set of standards. The example addresses the connections of the Common Core State Standards in Mathematics (CCSSM) (National Governors Association Center for Best Practices & Council of Chief State School Officers, 2010) to the proposed professional learning related to the Circumscribing Polygons Situation. The Common Core State Standards for mathematical content and mathematical practice mirror the attention given in various sets of standards both to what mathematics should be learned and to the mathematical processes in which students should engage. The Appendix of this Guidebook lists the CCSSM Standards for Mathematical Practice, one example of mathematical process standards. Questions (in bold italics in the following chart) that accompany the display of each CCSSM standard can be used in professional learning settings to extend work with specific standards.

Related Common Core Standards
CCSSM Standards for Mathematical Content
Grade 7 Geometry   **Draw, construct, and describe geometrical figures and describe the relationships between them.**   7.G.2.　　Draw (freehand, with ruler and protractor, and with technology) geometric shapes with given conditions. Focus on constructing triangles from three measures of angles or sides, noticing when the conditions determine a unique triangle, more than one triangle, or no triangle.   *In some cases, knowing the measures of three parts of a triangle is sufficient to determine a unique triangle. When there is insufficient information to specify a unique triangle (e.g., measures of only two parts are specified), relationships among some remaining pairs of parts of the triangle can still be determined. For example, in $\triangle ABC$, if AB and BC are known, what is the relationship between angle $\angle B$ and AC? (Note that the law of cosines can be used to determine this relationship.)*[3]
Grade 8 Geometry   **Understand congruence and similarity using physical models, transparencies, or geometry software.**   8.G.5.　　Use informal arguments to establish facts about the angle sum and exterior angle of triangles, about the angles created when parallel lines are cut by a transversal, and the angle–angle criterion for similarity of triangles.   *What prerequisite knowledge would a student need to develop any of these informal arguments?*
High School—Geometry   **Understand and apply theorems about circles.**   G.C.3.　　Construct the inscribed and circumscribed circles of a triangle, and prove properties of angles for a quadrilateral inscribed in a circle.   *How can inscribed triangles be used to prove properties of an inscribed quadrilateral?*   **Apply geometric concepts in modeling situations.**   G.MG.1.　　Use geometric shapes, their measures, and their properties to describe objects ….   *What are objects in students' surroundings that are easily described by geometric shapes?*

CCSSM Standards for Mathematical Practice[4]
SMP2. Reason abstractly and quantitatively.
SMP3. Construct viable arguments and critique the reasoning of others.
SMP6. Attend to precision.
SMP7. Look for and make use of structure.

SUGGESTIONS FOR USING THIS SITUATION

Facilitators may want to peruse what follows for an idea on how to use the Circumscribing Polygons Situation in their professional learning settings. The chart provides an Outline of Participant Activities and a summary of the Tools and projected Time required for implementation of the activities. Following the outline are Facilitator Notes that describe each of the suggested activities in greater detail.

Tools	Time
• Compass, protractor, straightedge • Handouts with tasks and tables • Copies of the Prompt • 3×5 note cards • Copies of the Foci and Commentaries • Activities could be adapted to use dynamic geometry tools such as Geometer's Sketchpad or Geogebra.	2–4 hours, can be done in a single session or across multiple sessions
Outline of Participant Activities (Details for these activities follow in the Facilitator Notes section.)	
Launch	Participants inscribe polygons in circles and circumscribe circles about polygons.
Activity 1.	After completing their table of conditions under which polygons are circumscribable, participants discuss characteristics of cyclic polygons.
Activity 2.	Participants read the Prompt and generate a variety of possible teacher responses to the question, "Can all polygons be circumscribed?" They then discuss the extent to which those responses are potentially helpful.
Activity 3.	After analyzing and explaining the mathematics in an assigned Focus, participants compare the approaches they used in Activity 1 to circumscribe polygons to the ideas in the first three Foci.
Reflect and assess learning	Participants connect the professional learning activities to their own classroom practice and assess what they learned from the session(s).

FACILITATOR NOTES

About the mathematics

This Situation offers opportunities to explore both the nature of circumscribed polygons and the logic and characteristics of biconditional statements. By proving many of the statements about circumscribed polygons, the mathematical topics can be nicely interwoven. However, there are advantages to beginning the exploration by considering each Focus separately. The Situation does consider concave and convex polygons, as well as polygons with nonsequential vertices.

Circumscribed polygons

Many polygons can be circumscribed. From a different perspective, one may say that many polygons can be inscribed in a circle. However, polygons that can be circumscribed must have special properties. Consider:

Some participants may know about circumscribing polygons, but may not be familiar with using the term *cyclic polygons* to classify polygons that can be circumscribed.

- All regular polygons can be circumscribed, but it is not the case that every circumscribed polygon is regular.

- All rectangles can be circumscribed, but not all quadrilaterals can be circumscribed.

- Quadrilaterals that can be circumscribed must have opposite angles that are supplementary.

- Parallelograms are special quadrilaterals with congruent opposite angles. Because rectangles are the only parallelograms with supplementary opposite angles, rectangles are the only parallelograms that can be circumscribed.

By looking at characteristics of circumscribed polygons, many important features of polygons and circles are made explicit.

- It is important to note that all points on a circle are equidistant from the center. Likewise, it is important to know that each point on the perpendicular bisector of a line segment is equidistant from the endpoints of the line segment.

- An angle inscribed in a circle cuts off (subtends) an arc on the circle whose measure is twice that of the inscribed angle. For example, consider an inscribed equilateral triangle. Each 60° angle cuts off $\frac{1}{3}$ of the circle, which has degree-measure 120 because the total number of degrees in a circle is 360.

In addition, the Situation addresses differences among concave and convex polygons, as well as defines and describes *n*-sided figures ($n \geq 4$) that do not have sequential vertices (e.g., the four-sided figure ▷◁).

Proof of biconditional statements

The idea that a quadrilateral can be circumscribed if and only if its opposite angles are supplementary is a biconditional statement. To be sure that this statement is true, one must prove two conditional statements: (a) If the quadrilateral is circumscribed, its opposite angles must be supplementary and (b) If the opposite angles are supplementary, the quadrilateral can be circumscribed. To understand the power of a biconditional statement it is important to consider statements that are not biconditional. For example, if a polygon is regular, it has a circumscribed circle. However, if a polygon has a circumscribed circle, it is not necessarily regular (e.g., all rectangles, isosceles trapezoids).

This Situation also demonstrates the power of posing a condition and proving it is false. To prove that a polygon with opposite, supplementary angles must be inscribable, one begins by posing the situation in which one vertex falls inside the circle and disproving that case. One proceeds by posing the situation in which the vertex falls outside the circle and disproving that case. Then it follows that the vertex must fall on the circle, completing the proof that such a polygon can always be circumscribed.

Launch

The goal of the launch is to explore characteristics of circumscribed polygons. Begin by looking at polygons from two different perspectives—inscribing polygons inside a given circle and circumscribing a given polygon with a circle. Suggested Handouts 1 and 2 appear at the end of this Guide.

Time

20–40 minutes

Launch Activity A

Distribute Handout 1, which has several large congruent circles. Ask participants to try to inscribe, using a straightedge, each of the following types of polygons in a drawn circle:

- triangle,

- quadrilateral,

- rectangle,

- regular hexagon,

- parallelogram,

- trapezoid,

- star, and

- a polygon of their own design.

Launch Activity B

Distribute Handout 2 with specific large polygons of approximately the same size and ask the participants to try to circumscribe the figure with a circle, using a compass and straightedge. The figures include:

- two rectangles of different dimensions,

- an irregular hexagon that cannot be circumscribed,

- a nonisosceles trapezoid,

- a scalene triangle,

- a concave polygon,

- an irregular quadrilateral that can be circumscribed, and

- an irregular quadrilateral that cannot be circumscribed.

Activity 1. Complete the chart in Handout 3 (the conditions under which polygons are circumscribable) and discuss characteristics of cyclic polygons.

Encourage participants to discuss the contents of their completed chart with others, drawing on their work with Handout 1 and Handout 2.

Time

20–30 minutes

Anticipated participant responses

Launch Activity A is probably easier than Launch Activity B because one needs to locate one of the infinitely many sets of appropriate vertices on the circle and then connect the vertices with straight line segments in Launch Activity A. Launch Activity B requires one to estimate (or perhaps actually construct) the center of the unique circle that will include all vertices. Participants may not remember how to find the center of the circle by constructing the perpendicular bisectors of two different sides of the polygon that can be circumscribed. It may not be clear that the perpendicular bisectors of each side of the polygon will intersect at a common point, which is the center of the circumscribed circle. If the perpendicular bisectors of each side of the polygon do not intersect in a common point, the polygon cannot be circumscribed. Participants should also notice that not all the assigned polygons can be circumscribed, and they should be encouraged to suggest which characteristics or properties prevent a polygon from being circumscribed. The chart created in answer to Handout 3 (see sample responses that appear at the end of this Guide) helps to summarize the information that the participants have found from the two activities in the Launch.

The task in Handout 3 asks participants to look both at the work they did with inscribing in Launch Activity A and with circumscribing in Launch Activity B. The purpose is to think about the relationship between inscribing and circumscribing. The two actions come from different perspectives and highlight different properties of cyclic polygons. Also, it is important to note that no concave polygons are cyclic. Four-sided planar figures with nonsequential vertices, rarely defined as polygons in most secondary curricula, can be circumscribed if their opposite angles are congruent.

Activity 2. Read the Prompt and generate a variety of possible teacher responses to the question, "Can all polygons be circumscribed?" Then sort the responses based upon how potentially helpful they might be.

Encourage participants to think about helpful, not helpful, and even incorrect teacher responses. Participants should produce teacher responses that might be given and not necessarily only those they would give to their students. This may relieve some pressure about offering a response that is only partially true and generate some useful discussion. Participants should work in small groups with each person writing an individual response on a separate note card so that they can be sorted into the three categories (i.e., incorrect, not helpful, and helpful). Each group should discuss members' responses and their classifications.

Time

20–30 minutes

Anticipated participant responses

When teachers respond to a student's question, such as "Can you circumscribe a circle about any polygon?", many factors need to be considered. An answer that is only partially true could be considered incorrect or not helpful. For example, the response, "Regular polygons can be inscribed," seems to have some key ideas but it is misleading. "Inscribed" suggests a relationship between two figures. The response names only one type of figure (regular polygons) and does not indicate the second type of figure.

Helpful responses might include:

- Some polygons can be circumscribed and some cannot. For example, all rectangles can be circumscribed. Can you think of some polygons that cannot be circumscribed?

- Hmm, let's see… (Teacher displays a parallelogram with an acute angle.) Can you circumscribe this parallelogram? Tell me why or why not.

- Let's think about this systematically. We know all triangles can be circumscribed. Can all squares? All rectangles? All parallelograms? All quadrilaterals?

Facilitating the activity

It is important to understand how different responses encourage or discourage mathematical thinking, mathematical understanding, and particularly specific mathematical practices.

Offer the following example.

It is quick and accurate to say, "No, not all polygons can be circumscribed. Here is a counterexample (teacher draws an irregular pentagon that clearly cannot be circumscribed)."

- Is the response correct? Helpful?

- What incomplete conceptions might be encouraged or supported by this response?

- What opportunities for mathematical reasoning or other practices are lost?

After participants work in their small groups, discuss responses that fall into each of the following categories: incorrect response, correct but unhelpful response, and correct and helpful response. Have groups give a rationale for where they placed a particular response.

The full group should discuss what is learned about circles and polygons as well as mathematical practices such as reasoning or defining. In addition, discuss what incomplete conceptions might be formed from various responses.

Activity 3. Analyze and explain the mathematics in an assigned Focus. After small group presentations, compare the approaches used in Activity 1 to circumscribe polygons to the ideas in the first three Foci.

Give copies of the entire Situation without Focus 4 to each participant and assign each group to look at one of the first three Foci. They should compare the ideas in the assigned Focus to the group discussion in Activity 1 and other ideas they generate as they consider their assigned Focus. Each group leads a brief discussion about the main points in their assigned Focus and entertains questions about the Focus.

Time

(45–60 minutes)

Anticipated participant responses

Participants might see similarities between ideas generated in the discussion and their assigned Focus. Participants may have ideas that are not addressed in their assigned Focus that may come up in another of the first three Foci, or in Focus 4, which is related to regular polygons and to reasoning. Participants may also suggest ideas that are not addressed in any of the four Foci but are ones they think could be valuable.

Facilitating the activity

Observations made during group work provide insights into how to relate the contents of the first three Foci. Challenge groups to prove their conjectures.

Key points about the first three Foci (Use these points to supplement group discussion.)

- The first three Foci each present both a claim and reasoning about that claim. Discussion about these Foci could play out along at least two pathways. The first pathway leads to a discussion about polygons, circles, and the act of circumscribing. The second pathway leads to a discussion about deductive reasoning and proving hypotheses. Both pathways are important in mathematics, and teacher should be able to integrate the two pathways as well as focus on them separately as is suggested in these activities.

- The Situation restricts discussion to Euclidean geometry in a plane.

- The Commentary explains that a cyclic polygon is defined as one that can be circumscribed by a circle and notes that the relationships among the angles, sides, and the perpendicular bisectors of the sides in a polygon determine whether a polygon can be circumscribed. The Foci that follow describe classes of cyclic polygons.

- Focus 1 explains why all triangles can be circumscribed and how this generalization is important for exploring other polygons. The Focus introduces the power of finding a perpendicular bisector for each side and defines the circumcenter of a triangle. The Focus also discusses two ways to show that all triangles can be circumscribed; it offers a geometric approach and an algebraic approach.

- Focus 2 addresses convex quadrilaterals and explains that a convex quadrilateral is cyclic if and only if the opposite angles are supplementary. The focus illustrates the meaning of *if and only if* by explaining that the two conditions of the quadrilateral being cyclic and the quadrilateral having opposite supplementary angles are equivalent conditions. The emphasis on deductive reasoning leads to the revelation that a rectangle is the only parallelogram that can be circumscribed.

- Focus 3 takes on unusual four-sided figures that often are ignored when exploring planar figures. The Focus explains why concave quadrilaterals cannot be circumscribed and contrasts convex and concave quadrilaterals. In addition, nonsequential quadrilaterals are defined and conditions for circumscribing such quadrilaterals are shown. There is an emphasis on working with precise definitions and deductive reasoning.

Key points for discussion

- Engage participants in a reflective discussion about the two pathways embedded in the first three Foci. All three Foci addressed both geometrical ideas and proving geometric hypotheses.

  ○ Which dimension (i.e., geometry or proof) was most comfortable for you? Why?

  ○ What geometrical ideas were new or less familiar to you?

  ○ What deductive arguments were new or less familiar to you?

  ○ How were biconditional statements important in this Situation?

Reflect and assess learning

The purpose is for participants to connect the session's activities to their own classroom practice, and to assess what participants learned from the session. Specific activities depend on the setting for the professional learning session, that is, whether it is a stand-alone session or part of an ongoing series of sessions.

Time

40–60 minutes

Suggested assessment activities

- Ask participants to prove that every planar regular polygon is cyclic, but not every cyclic polygon is regular. Participants should work individually and then discuss their progress with their group. Then the group should compare its discussion to the proof in Focus 4. It may not be appropriate for every group to develop a rigorous proof within the time provided, but an attempt to prove the statement will help the participants appreciate the arguments presented in Focus 4 and the Postcommentary. It is important that all participants understand the power of an *if and only if* statement and the fact that one counterexample can disprove a statement. (However, it is also very important to note that one example does not prove a statement.)

  ○ Focus 4 offers a proof that every planar regular polygon is cyclic, but not every cyclic polygon is regular. It emphasizes that the converse of every true statement is not necessarily true.

  ○ The Postcommentary explains the power of a biconditional statement as shown in Focus 2 and compares it to the statement proved in Focus 4. It also provides a proof for the biconditional statement used in Focus 1.

- Ask participants to write about how what they have learned will enhance their instruction. You may want to tailor the assessment to specific grade bands (i.e., middle school or high school) or courses (i.e., geometry, algebra, advanced mathematics).

Reflection Questions

1. How have the activities and our discussion enhanced your ideas about these topics?

2. How do the following topics generalize to mathematics instruction in general?

    a. *Definitions.* Definitions influence deductive reasoning. The Foci bring up interesting definitions for trapezoids, concave polygons, and nonsequential quadrilaterals. How does the choice of definition used affect the reasoning needed?

    b. *Geometric constructions.* When constructing a perpendicular bisector of the side of a polygon one needs to emphasize not only how to construct the bisector but also the properties of the bisector. How can angle bisectors be used to claim concave quadrilaterals cannot be circumscribed?

    c. *Proof and proving.* Helping students understand the difference between proving a statement and providing a collection of examples is critical in mathematics. How would the presentation of examples and the justifying be integrated in this Situation?

    d. *Examples and nonexamples.* The initial activities of inscribing and circumscribing polygons help to generate a variety of examples of polygons that could be circumscribed. Hopefully they also caused participants to create some nonexamples and to develop some hypotheses about what classes of polygons could and could not be circumscribed.

3. Ask participants to reflect on their experiences in generating the polygons and circles and then identify difficulties their students might have with circumscribing polygons or finding the circumcenter of cyclic polygons.

4. Ask participants to reflect on the biconditional statements and proofs in their work with the Situation and anticipate their students' thinking about the nature of biconditional proofs. What might be the impact of the integration of learning about geometric figures and deductive reasoning?

5. Ask participants to reflect on which of the Standards for Mathematical Practice they were engaged in during the session. They may identify many of the Standards for Mathematical Practice. The following were part of the design of these activities.

    a. *Reason abstractly and quantitatively.* The activities ask participants to generate examples and to provide arguments for their decisions. They were asked to generate subgroups of geometric figures and to create categories depending on the features of the geometric figures.

b. *Construct viable arguments and critique the reasoning of others.* The Foci emphasized the nature of biconditional statements. Hypotheses were supported with deductive reasoning. The Foci noted that the converse of statements cannot be assumed to be true based solely on the original conditional statement being true.

c. *Attend to precision.* The activities used precise definitions and showed how to use definitions to answer questions and generate both examples and nonexamples.

d. *Look for and make use of structure.* Focus 2 showed how a geometric perspective and an algebraic perspective provide two very different ways of addressing a problem. Comparisons between inscribing and circumscribing were made. Classes of geometric figures were characterized.

RESOURCE

Aichele, D. B., & Wolfe, J. (2008). *Geometric structures: An inquiry-based approach for prospective elementary and middle school teachers.* Upper Saddle River, NJ: Pearson. [Chapter 9 of Aichele and Wolfe (2008) can be a helpful resource for work with elementary and middle school teachers.]

NOTES

1. The Circumscribing Polygons Situation is one of the Situations presented in *Mathematical Understanding for Secondary Teaching: A Framework and Classroom-Based Situations* (Heid & Wilson, 2015).

2. This Situation appears on pp. 365–375 of Heid and Wilson (2015). It is reprinted with permission.

3. For each standard, we have suggested questions (those in italicized, boldface type) that might serve as follow-up questions to address the mathematics of the standard.

4. See the Appendix for the list of the Standards for Mathematical Practice in the Common Core State Standards for Mathematics (National Governors Association Center for Best Practices and Council of Chief State School Officers, 2010).

REFERENCES

Heid, M. K. & Wilson, P. W. (with G. W. Blume). (Eds.). (2015). *Mathematical understanding for secondary teaching: A framework and classroom-based situations.* Charlotte, NC: Information Age.

National Governors Association Center for Best Practices & Council of Chief State School Officers. (2010). *Common core state standards for mathematics.* Washington, DC: Authors. Retrieved from www.corestandards.org/uploads/Math_Standards1.pdf

CIRCUMSCRIBING POLYGONS HANDOUT 1

Inscribe one polygon in each of the given circles. Your polygons should include at least a triangle, a quadrilateral, a rectangle, a regular hexagon, a parallelogram, a trapezoid, a star, and a polygon of your own design.

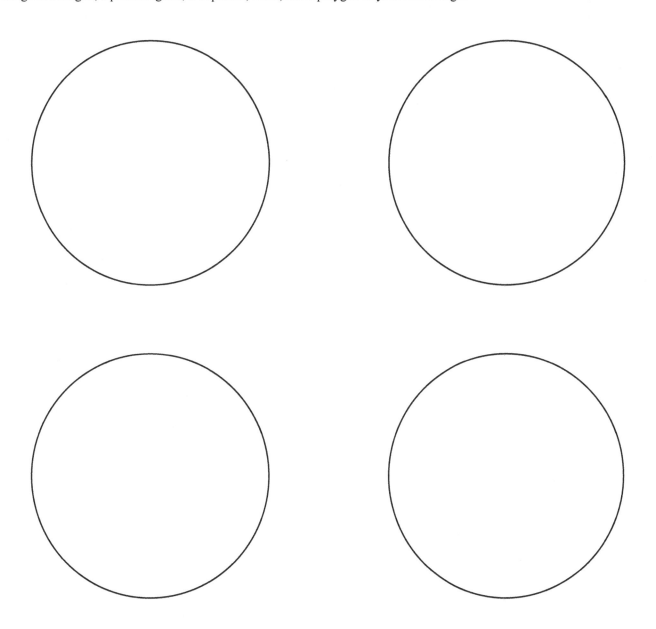

CIRCUMSCRIBING POLYGONS HANDOUT 1 (CONTINUED)

CIRCUMSCRIBING POLYGONS HANDOUT 2

For each of the following figures, circumscribe the figure or explain why it cannot be circumscribed:

CIRCUMSCRIBING POLYGONS HANDOUT 2 (CONTINUED)

For each of the following figures, circumscribe the figure or explain why it cannot be circumscribed:

CIRCUMSCRIBING POLYGONS HANDOUT 3

Complete the following chart:

Figure	Can it be circumscribed?	Are there special conditions?
Triangle		
Scalene triangle		
Rectangle		
Parallelogram		
Trapezoid		
Regular polygon		
Concave polygon		
Irregular polygon		

CIRCUMSCRIBING POLYGONS HANDOUT 3 WITH RESPONSES

Complete the following chart:

Figure	Can it be circumscribed?	Are there special conditions?
Triangle	*Yes*	*Any triangle is cyclic.*
Scalene triangle	*Yes*	*Any triangle is cyclic.*
Rectangle	*Yes*	*Any convex quadrilateral with opposite supplementary angles is cyclic.*
Parallelogram	*Maybe*	*Only rectangles* *[It is important to note that the ONLY parallelograms that are cyclic are rectangles.]*
Trapezoid	*Maybe*	*Only isosceles trapezoids* *[Definitions of trapezoids may vary.]*
Regular polygon	*Yes*	*Regular polygons have congruent angles and sides.*
Concave polygon	*No*	*Cannot be circumscribed* *[Note the difference between a concave polygon and a figure with nonsequential sides (e.g., ▷◁).]*
Irregular polygon	*Maybe*	*Not all cyclic polygons are regular.* *Not all irregular polygons are cyclic.*

CHAPTER 6

FACILITATOR'S GUIDE
FOR
CALCULATION OF SINE[1]

Situation 35 From the MACMTL–CPTM Situations Project

Steven S. Viktora and James W. Wilson

Calculation of values of the sine function, once presented as an exercise in interpolation in a table of values, and now presented as an exercise in using a calculator, is situated richly in a range of mathematical ideas from graphical interpolation to circular functions and tangent-line approximations. Myriad potential connections across representations and definitions make the calculation of trigonometric values fertile ground for consolidating students' understanding of essential mathematical ideas. This chapter offers ways that teachers can explore an often taken-for-granted concept.

Facilitator's Guidebook for Use of Mathematics Situations in Professional Learning,
pages 115–134.
Copyright © 2018 by Information Age Publishing

OVERVIEW

Facilitators can scan the following overview to quickly get a sense of the mathematics involved in the proposed professional learning setting. The sections of the overview deal with the mathematics of the Calculation of Sine Situation, why these mathematical ideas might be important for participants, the learning goals for participants, and mathematical ideas central to the proposed professional learning sessions.

Situation	Relevance
A class demonstration and discussion on values of trigonometric functions for special angles such as 30° and 45° led a student to ask, "How do you calculate the approximation for sin 32°?"	• The calculation of approximate values of trigonometric functions is a fundamental conceptual issue. Although students have access to programmed keys on technology devices for use when solving problems, discussion of ways to calculate the approximations from various measurements and other information can lead to better understanding. • Although the question was about a specific value of the sine function, it generalizes easily to all trigonometric functions and draws on core understanding of the meanings of these trigonometric values. • Estimation and approximation are important mathematics processes.
Goals	Key mathematical ideas
• Explore and contrast approximations of trigonometric function values by various means. • Examine concepts of approximation, measure, and precision. • Consider how and when to address this issue with students.	• Trigonometric functions for angles with degree measures less than 90 can be defined in terms of right triangles. • The unit circle allows one to define trigonometric functions for any real-valued angle measures. • Ratios and lines offer different approaches to estimating and approximating values of trigonometric functions.

The Situation under consideration in this chapter follows (highlighted with a gray background).[2]

CALCULATION OF SINE

Situation 35 From the MACMTL–CPTM Situations Project

Patricia S. Wilson, Heather Johnson, Jeanne Shimizu, Evan McClintock, Rose Mary Zbiek, M. Kathleen Heid, Maureen Grady, and Svetlana Konnova

PROMPT

After completing a discussion on special right triangles (30°–60°–90° and 45°–45°–90°), the teacher showed students how to calculate the sine of various angles using a calculator.

A student then asked, "How could I calculate sin(32°) if I do not have a calculator?"

COMMENTARY

The set of Foci provide interpretations of sine as a ratio and sine as a function, using graphical and geometric representations. The first three Foci highlight $\sin(\theta)$ as a ratio, appealing to the law of sines, right-triangle trigonometry, and unit-circle trigonometry. The next three Foci highlight $\sin(x)$ as a function and use tangent and secant lines as well as polynomials to approximate $\sin(x)$.

The question of "how good" an approximation one gets using secants, tangents, or Taylor polynomials depends on the size of the x-interval, the order of the highest derivative, and the function(s) in question. These challenging problems are usually taken up in courses on mathematical analysis and numerical analysis.

MATHEMATICAL FOCI

Mathematical Focus 1

Ratios of lengths of sides of right triangles can be used to compute and approximate trigonometric function values.

A ratio of measures of sides of a right triangle with an acute angle of measure $x°$ can be used to approximate $\sin(x)$. $\mathrm{Sin}(x)$ can be approximated by sketching a 32°–58°–90° right triangle with a protractor or with dynamical geometry software, measuring the length of the hypotenuse and leg opposite the 32° angle, and computing the sine ratio (see Figure 41.1).

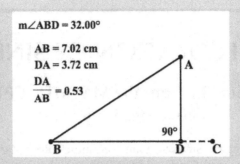

FIGURE 41.1. Right triangle ABD with a 32° angle.

Hence, $\sin(32°) \approx 0.53$.

Mathematical Focus 2

Coordinates of points on the unit circle represent ordered pairs of the form (cos(θ), sin(θ)) that can be used to approximate trigonometric values.

The unit circle is the locus of all points one unit from the origin $(0, 0)$. The equation for a circle with radius 1 centered at the origin is $x^2 + y^2 = 1$. Consider the angle θ in standard position formed by the x-axis and a ray from the origin through a point A on the unit circle. Then, $\cos(\theta) = \dfrac{x}{1}$ and $\sin(\theta) = \dfrac{y}{1}$. Hence, the coordinates of A are $(\cos(\theta), \sin(\theta))$, and another equation for a circle with radius 1 centered at the origin is $(\cos(\theta))^2 + (\sin(\theta))^2 = 1$.

Let A be positioned on the unit circle so that $\angle ABD$ has degree-measure 32° (see Figure 41.2). Then, the signed length of segment AD is equal to $\sin(32°)$. The signed length of segment AD is approximately 0.53, and so $\sin(32°) \approx 0.53$.

Mathematical Focus 3

The law of sines can be used to compute and approximate the sine function value through the measurement of geometric constructions.

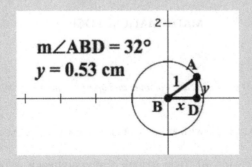

FIGURE 41.2. Right triangle ABD with a 32° angle on a unit circle.

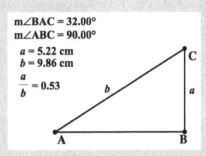

FIGURE 41.3. Using the law of sines
and sin(90°) to calculate sin(32°).

The law of sines applies to any triangle in a plane. Consider triangle ABC, with sidelengths a, b, and c for $\overline{BC}$, $\overline{AC}$, and

$\overline{AB}$, respectively. The law of sines states:

$$\frac{a}{\sin(A)} = \frac{b}{\sin(B)} = \frac{c}{\sin(C)}.$$

Sin(32°) can be approximated by sketching any triangle the degree-measure of one of whose angles is 32° and the degree-

measure of another of whose angles has a known sine value (e.g., 30°, 45°, 60°, or 90°).

For example, a triangle can be sketched (with dynamical geometry software) with m∠A = 32° and m∠B = 90° (see Fig-

ure 41.3). Using the measure a and the measure b (the length of the side opposite the 90° angle), sin(32°) can be calculated

using the law of sines.

$$\frac{a}{\sin(32°)} = \frac{b}{\sin(90°)}$$

Because sin(90°) = 1, $\sin(32°) = \frac{a}{b}$.

Hence, sin(32°) ≈ 0.53.

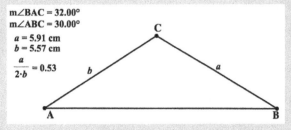

FIGURE 41.4. Using the law of sines and sin(30°) to
calculate sin(32°).

For another example, a triangle can be sketched (with dynamical geometry software) with m∠A = 32° and m∠B = 30° (see Figure 41.4). Using the measure a and the measure b (the length of the side opposite the 30° angle), sin(32°) can be calculated using the law of sines.

$$\frac{a}{\sin(32°)} = \frac{b}{\sin(30°)}$$

Because $\sin(30°) = \frac{1}{2}$, $\sin(32°) = \frac{a}{2b}$. Hence, sin(32°) ≈ 0.53.

Mathematical Focus 4

A continuous function, such as f(x) = sin(x), can be represented locally by a linear function and that linear function can be used to approximate local values of the original function.

The function $f(x) = \sin(x)$ is not a linear function; however, linear functions can be used to approximate nonlinear functions over sufficiently small intervals.

Measuring angles in radians, 180° is equivalent to π radians. Therefore:

$$30° \text{ is equivalent to } \frac{30\pi}{180} = \frac{\pi}{6} \text{ , or } 0.5236 \text{ radians,}$$

$$32° \text{ is equivalent to } \frac{32\pi}{180} = \frac{8\pi}{45} \text{ , or } 0.5585 \text{ radians, and}$$

$$45° \text{ is equivalent to } \frac{45\pi}{180} = \frac{\pi}{4} \text{ , or } 0.7854 \text{ radians.}$$

Figure 41.5 shows the graph of the function $f(x) = \sin(x)$ and the graph of the secant line $\overleftrightarrow{AB}$, where the coordinates of A are $\left(\frac{\pi}{6}, \sin\left(\frac{\pi}{6}\right)\right) = \left(\frac{\pi}{6}, \frac{1}{2}\right) = \left(\frac{\pi}{6}, 0.5\right)$ and the coordinates of B are $\left(\frac{\pi}{4}, \sin\left(\frac{\pi}{4}\right)\right) = \left(\frac{\pi}{4}, \frac{\sqrt{2}}{2}\right) = \left(\frac{\pi}{4}, 0.7071\right)$. Because the func-

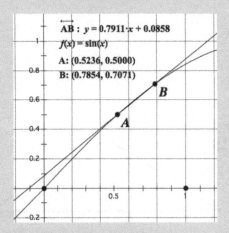

FIGURE 41.5. Using a secant line to estimate sin(32°).

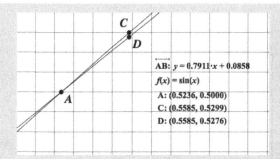

FIGURE 41.6.   Zooming in on the estimate of
sin(32°) using a secant line.

tion $f(x) = \sin(x)$ is approximately linear between points A and B, the values of the points on the secant line, $\overrightarrow{AB}$, provide

reasonable approximations for the values of $f(x) = \sin(x)$ between points A and B (see Figure 41.5). Because $\sin(x)$ is concave

down in the x-interval $\left\langle \dfrac{\pi}{6}, \dfrac{\pi}{4} \right\rangle$, the estimate for sin(32°) will be an underestimate.

In Figure 41.6, point D on secant line $\overrightarrow{AB}$ with coordinates (0.5585, 0.5276) provides a reasonable approximation for the

location of point C on $f(x) = \sin(x)$ with coordinates (0.5585, sin(0.5585)). Therefore, sin(32°) ≈ 0.5276.

An approximation for sin(32°) can also be found by using the equation for secant line $\overrightarrow{AB}$. Because secant line $\overrightarrow{AB}$

passes through the points $\left(\dfrac{\pi}{6}, \sin\left(\dfrac{\pi}{6}\right) \right) \approx (0.5236, 0.5)$ and $\left(\dfrac{\pi}{4}, \sin\left(\dfrac{\pi}{4}\right) \right) \approx (0.7854, 0.7071)$, its equation can be approximated

as follows:

$$y - 0.5 = \frac{0.7071 - 0.5}{0.7854 - 0.5236}(x - 0.5236)$$
$$y = 0.7911(x - 0.5236) + 0.5$$

When $x = 0.5585$, $y = 0.5276$. Therefore, sin(32°) ≈ 0.5276.

Mathematical Focus 5

Given a differentiable function and a line tangent to the function at a point, values on the tangent line will approximate values of the

function near the point of tangency.

Because the function $f(x) = \sin(x)$ is differentiable, given a point $(a, \sin(a))$ on $f(x) = \sin(x)$, the line tangent to $f(x) = \sin(x)$

at $(a, \sin(a))$ can be used to approximate $(a + dx, \sin(a + dx))$ at a nearby point with x-coordinate $a + dx$. When dx is small,

the y-value of the tangent line at the point with x-coordinate $a + dx$ will be very close to the value of $\sin(a + dx)$. Using radian

measure, 32° is equivalent to $\dfrac{32\pi}{180} = \dfrac{8\pi}{45}$, or 0.5585 radians.

Consider a geometric interpretation of differentials dx and dy and their relation to Δx and Δy, where a tangent line is used to

approximate $f(x)$ near a given value (see Figure 41.7).

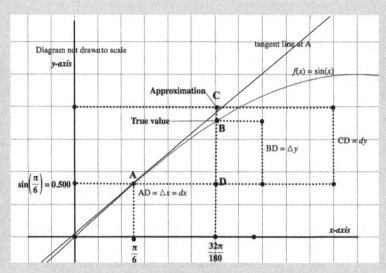

FIGURE 41.7. A geometric interpretation of differentials to estimate sin(32°).

$$f'(x) \approx \frac{\Delta y}{\Delta x} \rightarrow \Delta y \approx (\Delta x) f'(x)$$

Because $\left(a,\ f(a)\right) = \left(\frac{\pi}{6},\ \sin\left(\frac{\pi}{6}\right)\right)$ and $f'(x) = \cos(x)$,

$$\Delta y \approx (\Delta x) f'(x)$$
$$\Rightarrow \sin\left(\frac{32\pi}{180}\right) - \sin\left(\frac{\pi}{6}\right) \approx \left(\frac{32\pi}{180} - \frac{\pi}{6}\right)\cos\left(\frac{\pi}{6}\right)$$
$$\Rightarrow \sin\left(\frac{32\pi}{180}\right) - \sin\left(\frac{\pi}{6}\right) \approx 0.0302$$
$$\Rightarrow \sin\left(\frac{32\pi}{180}\right) \approx 0.0302 + \sin\left(\frac{\pi}{6}\right)$$
$$\Rightarrow \sin\left(\frac{32\pi}{180}\right) \approx 0.5302 .$$

Mathematical Focus 6

Mathematical theory involving Taylor series provides a definition of the sine function based on the foundations of the real number system, independent of any geometric considerations.

The sine function could be defined using an infinite series. The following identity holds for all real numbers x, with angles measured in radians:

$$\sin x = x - \frac{x^3}{3!} + \frac{x^5}{5!} - \frac{x^7}{7!} + \ldots = \sum_{n=0}^{\infty} \frac{(-1)^n x^{2n+1}}{(2n+1)!}$$

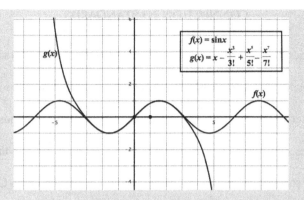

FIGURE 41.8.   A Taylor polynomial approximating the sine function.

The sine function is closely approximated by its Taylor polynomial of degree 7 for a full cycle centered on the origin, $-\pi \leq x \leq \pi$ (see Figure 41.8).

POSTCOMMENTARY

Although the methods shown in the Foci differ in the use of ratios versus the use of lines as approximation tools for the sine function, each of the methods involves approximations. The ratio methods depend on a definition of the trigonometric functions and therefore are not generalizable to other types of functions, whereas the line methods depend on characteristics of continuous functions and therefore can be used for a wider range of functions.

CONNECTION TO STANDARDS

Learning experiences in mathematics for teachers are well positioned when teachers are aware of connections to the standards

for which their students are accountable. These standards differ across states and provinces as well as across countries, although

there are commonalities across different sets of standards. Although this document cannot possibly address all the existing sets

of standards across states, provinces, and countries, an example follows that illustrates how the proposed professional learning

might address one particular set of standards. The example addresses the connections of the Common Core State Standards for

Mathematics (CCSSM) (National Governors Association Center for Best Practices & Council of Chief State School Officers,

2010) to the proposed professional learning related to the Calculation of Sine Situation. The Common Core State Standards for

mathematical content and mathematical practice mirror the attention given in various sets of standards both to what mathemat-

ics should be learned and to the mathematical processes in which students should engage. The Appendix of this Guidebook lists

the CCSSM Standards for Mathematical Practice, one example of mathematical process standards. Questions (in bold italics

in the following table) that accompany the display of each CCSSM standard can be used in professional learning settings to

extend work with specific standards.

Related Common Core Standards
CCSSM Standards for Mathematical Content

High School—Functions	
Interpret functions that arise in applications in terms of the context.	
F-IF.6.	Calculate and interpret the average rate of change of a function (presented symbolically or as a table) over a specified interval. Estimate the rate of change from a graph. *How might one approximate sin(50°) using the known values of sin(45°) and sin(60°)?*
Analyze functions using different representations.	
F-IF.7e.	Graph exponential and logarithmic functions, showing intercepts and end behavior, and trigonometric functions, showing period, midline, and amplitude. *Consider the claim, "If a > 32, then sin(a°) > sin(32°). How might one use the graph of the sine function to assess the truth of this statement?*
Extend the domain of trigonometric functions using the unit circle.	
F-TF.1.	Understand radian measure of an angle as the length of the arc on the unit circle subtended by the angle. *Suppose one knows the length of an arc of a circle with radius r, how can one find the measure of the angle that subtends that arc?*
F-TF.2.	Explain how the unit circle in the coordinate plane enables the extension of trigonometric functions to all real numbers, interpreted as radian measures of angles traversed counterclockwise around the unit circle. *Explain how one could use the unit circle to find the values of sin(x + π) and sin(x + 2π) if the value of sin(x) is known.*
F-TF.3. (+)[3]	Use special triangles to determine geometrically the values of sine, cosine, tangent for $\pi/3$, $\pi/4$ and $\pi/6$, and use the unit circle to express the values of sine, cosine, and tangent for $\pi - x$, $\pi + x$, and $2\pi - x$ in terms of their values for x, where x is any real number. *Carry out the action required in this standard and use the relationship between right triangles and the unit circle to explain why your strategy works.*

High School—Geometry	
Apply trigonometry to general triangles.	
G-SRT.10.(+)	Prove the Laws of Sines and Cosines and use them to solve problems. ***The law of sines involves the measures of angles and sides in a triangle. Can the law of sines be used to approximate the value of sin(x) when x > 180°?***
G-SRT.11.(+)	Understand and apply the Law of Sines and the Law of Cosines to find unknown measurements in right and non-right triangles (e.g., surveying problems, resultant forces). ***Suppose one wants to use the law of sines to find sin(a), where a, b, and c are the side lengths opposite angles A, B, and C, respectively, in triangle ABC. How might one obtain the three other values needed to find the value of sin(a)?***
CCSSM Standards for Mathematical Practice[4]	
SMP2.	Reason abstractly and quantitatively.
SMP3.	Construct viable arguments and critique the reasoning of others.
SMP6.	Attend to precision.

SUGGESTIONS FOR USING THIS SITUATION

Facilitators may want to peruse what follows for an idea about how to use the Calculation of Sine Situation in their professional learning settings. The chart provides an Outline of Participant Activities and a summary of the Tools and projected Time required for implementation of those activities. Following the chart are Facilitator Notes that describe each of the suggested activities in greater detail.

Tools	Time
• Calculators and/or other computing technologies (e.g., spreadsheet, smartphone applet) • Graphing utility (e.g., graphing calculator, dynamic geometry system, online computer algebra system) • Poster paper, markers • Copies of the Prompt (separate from the Foci) • Copies of the Foci • On poster board, copies of a 32°–58°–90° right triangle that can be physically measured	2–3 hours, done in a single session or across multiple sessions
Outline of Participant Activities (Details for these activities follow in the Facilitator Notes section.)	
Launch	Participants generate ideas about ways that the approximate value of a trigonometric function can be determined without reference to a table of function values or calculation technology.
Activity 1.	Using the approaches generated in the Launch (and perhaps others), participants describe the steps required for various methods for calculating approximations of the values of trigonometric functions.
Activity 2.	Participants discuss the relationship between right-triangle trigonometry and unit-circle trigonometry.
Activity 3.	Participants describe the mathematical basis for each Focus and compare and contrast the approximation approaches in the Foci.
Activity 4.	Participants examine and contrast ideas from the Foci and engage in group discussions.
Reflect and assess learning	Participants describe how the activities extended their knowledge of approximation of trigonometric ratios and what ideas they might develop with their students.

FACILITATOR NOTES

About the mathematics

The interpretations of sine as a ratio and sine as a function can be contrasted using graphical, geometric, and symbolic representations. Discussions of various means of calculating the approximations of the sine values can lead to better understanding of connections to various areas of mathematics and applied problems. Approximation, measure, and precision are mathematical concepts worthy of discussion. The following are some informative resources pertinent to the Calculation of Sine Situation: Gelfand and Saul (2001), Maor (1998), and Usiskin, Peressini, Marchisotto, and Stanley (2003).

Launch

Time

15–20 minutes

The Launch focuses participants on naming various approaches they have used to approximate trigonometric functions. Have participants read the Prompt and ask them how they might calculate sin(32°) if they had neither a calculator nor a table of trigonometric function values available. The following are approaches addressed in the Situation's Foci:

- using right triangle definition of sine ratio,

- using unit circle definitions of sine ratio and sine function,

- using law of sines and connection to the subtended angle theorem from geometry,

- using secants of a continuous function over a small region,

- using tangent of a differentiable function at the point of tangency, and

- using polynomials from Taylor expansions.

Participants may not generate all of these, but subsequent activities will address these approaches in addition to those they produce. The extent and depth of discussion of the approaches from the six Foci need not be considered of equal importance to a particular group of participants. Participants should not fully develop the approaches they identify; they should merely name them at this time. Participants will have an opportunity to explore the approaches more fully in Activity 3.

Activity 1. Using the approaches generated in the Launch (and perhaps others), describe the steps required for various methods for calculating approximations of the values of trigonometric functions.

This activity might be completed either in small groups or as a whole-group orientation and discussion.

Time

7–10 minutes

Anticipated participant responses

Participants may need guidance about exactly what they are to do and how far to carry out their explanations as they outline their proposed strategies. For example, participants might offer suggestions such as "use the unit circle," but in this activity they should be more precise than that by describing the actions they would take to develop an approximation using the unit circle. In this case participants might say, "First construct a circle of radius 1 centered at the origin. Then embed a 32°–58°–90° right triangle in the circle, with the 32° angle in standard position and the hypotenuse drawn from the origin to a point on the circle. Then sin(32°) is the y-coordinate of that point." Participants might want to continue by completing their calculations to obtain a numerical approximation for that y-coordinate, but in this activity they should describe, but not carry out, the actions and calculations. At this point, understanding the mathematical relationships is more important than carrying out the actual calculations.

Facilitating the activity

Let the participants discuss but not carry out any calculations at this point. Encourage participants to think about trigonometric values from the approaches of right triangles and the unit circle. This might be done in small groups or as a whole-group session. The facilitator may need to emphasize the role of estimation and approximation as underlying processes in this trigonometric exploration.

The facilitator may need to convey that use of technology is not the focus of this Situation, and each of these approximation methods could be carried out without digital technology. Digital technology tools are convenient for drawing images and avoiding physical measurement. The calculator is a useful tool for calculating the ratios. A graphing technology tool facilitates the graphing of functions. None of these tool uses necessarily addresses the underlying concepts. They, in fact, may be a drawback because users (both teachers and students) tend to associate the precision embedded in technology tools with a belief that technology-generated values are "exact" measures.

Activity 2. Discuss the relationship between right-triangle trigonometry and unit-circle trigonometry.

Time

10 minutes

Facilitating the activity

Encourage participants to describe how finding trigonometric values using the approaches of right triangles and the unit circle are related. This might be done in small groups or as a whole-group session.

Activity 3. Describe the mathematical basis for each Focus and compare and contrast the Foci.

Time

1.5–2 hours

Facilitating the activity

This activity may be structured in a number of ways, depending on the mathematical understanding of the participants, their instructional level (middle school or high school), and the amount of time available. The amount of emphasis on particular Foci will also depend on participants' understanding and background.

Option 1. All groups analyze all Foci.

Give the following instructions to groups of participants:

- Describe the mathematical background needed to understand each Focus, and

- Compare and contrast the Foci on a range of characteristics of your choosing.

Groups should be prepared to present any of the Foci. If participants do not address comparisons/contrasts such as representation, ease of use, precision, and required mathematics background, consider asking them to compare and contrast the Foci with respect to each of these or other characteristics. Also (or alternatively) ask them to rank the Foci in terms of which approach they find most compelling or transparent for students whom they teach. (Note: Asking participants to rank items often produces richer discussion than simply asking them to discuss or analyze each one.) Also consider asking how each Focus advances the participant's own (and perhaps their students') understanding of the approximation of trigonometric ratios.

Option 2: Each group analyzes only one Focus or a subset of Foci.

Give the following instructions to groups of participants:

- Describe the underlying mathematics in your assigned Focus or Foci, and

- Create a 3-minute to 7-minute explanation of the method used in your Focus (or Foci) to calculate sine values. Include an explanation of why this approach produces a reasonable approximation of sin(32°).

Assign a different Focus or subset of the Foci to each group. In doing so, take into consideration what you know about their mathematics background and analysis of the approximation of trigonometric ratios. Each group should prepare a presentation about its Focus: how the approach is carried out and why the approach produces a reasonable approximation for sin(32°). While listening to other groups' presentations about their Foci, participants should be asked to compare and contrast their Focus with that of others and also consider how each Focus advances their own understanding of the approximation of trigonometric ratios.

Key points about the Foci

- Focus 1 addresses using right triangle definitions of the trigonometric ratios. The participants should construct and label appropriate images.

- Focus 2 uses the unit circle definitions of trigonometric ratios and functions, considering a coordinate plane and the point (cos θ, sin θ) on the unit circle. The participants should use, construct, and label appropriate images.

- Focus 3 is based on the law of sines, using:

  - a right triangle with a 32° angle, using the known value of sin 90°, and

  - a 32°–30°–118° triangle, using the known value of sin 30°.

- Focus 4 considers the continuous function $f(x) = \sin x$ and then approximates the value of sin 32° by either considering the secant between two known sine values that are relatively close to each other (such as sin 30° and sin 45°) and considering the secant line between the two points or by using the equation of the secant line. Participants might notice that this method is equivalent to *linear interpolation,* and they should be encouraged to explain this equivalence.

- Focus 5 considers $f(x) = \sin x$ as a continuous differentiable function and uses the tangent line to the function at the point to approximate the value of the function.

- Focus 6 uses the Taylor series definition of the sine function to generate polynomial approximations of the sine values.

- The Postcommentary points out that the ratio approaches are limited in generality to the trigonometric functions, whereas the line methods generalize to other classes of functions.

Additional optional explorations

- Consider exploring some or all of the methods to estimate the values of other trigonometric ratios such as cos 32° or tan 32°.

- Suppose that you were given that sin θ ≈ 0.3010. How would you find θ?

- Show how to generate the Taylor series identity in Focus 6.

- Identify other derivations and theorems that may be needed to understand the six Foci.

Activity 4. Examine and contrast ideas from the Foci and engage in group discussions.

Time

30–45 minutes

Key points for discussion

- Contrast the right-triangle definition of trigonometric ratios and the unit-circle definitions on the basis of the angles

  to which they apply.

- Discuss the law of sines and its derivation from the inscribed-angle theorem from geometry.

- Engage participants in a reflective discussion about their learning from this activity:

  ○ Which Foci have you considered before?

  ○ Which were new to you?

  ○ Which ones were most compelling?

  ○ What new insights did you gain, if any?

  ○ What confusion remains about the various approaches represented by the different Foci?

  ○ What is missing (e.g., historical context, applications)?

Alternative methods

One alternative method is based on using the theorem that states that for small θ in radian measure, $\sin(\theta) \approx \theta$, and cos

$\theta \approx 1$. Rewriting $\sin(32°)$ using the formula for $\sin(a + b)$,

$$\sin(32°) = \sin(30° + 2°) = \sin(30°)\cos(2°) + \sin(2°)\cos(30°).$$

Substituting 1 for $\cos(2°)$ and $\pi/90$ for $\sin(2°)$,

$$\sin(32°) \approx \frac{1}{2} \cdot 1 + \frac{\pi}{90} \cdot \frac{\sqrt{3}}{2} \approx 0.5302$$

The Taylor Series exploration could be done with generating the series centered at $\frac{\pi}{6}$ rather than at 0. This would give

the first four terms of the Taylor Series as

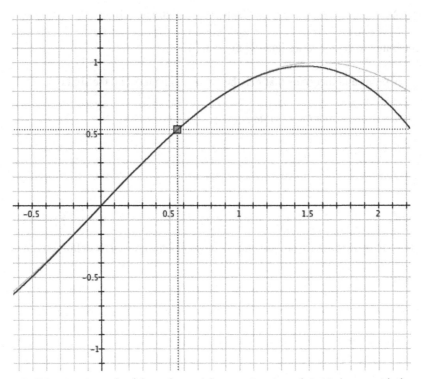

FIGURE 6.1. A graph of the polynomial approximation of sin(x) shown with the graph of sin(x) and the point with the function value $\frac{32\pi}{180}$.

$$\sin(x) = \frac{1}{2} + \frac{\sqrt{3}}{2}\left(x - \frac{\pi}{6}\right) - \frac{1}{4}\left(x - \frac{\pi}{6}\right)^2 - \frac{1}{4\sqrt{3}}\left(x - \frac{\pi}{6}\right)^3$$

The graph of the polynomial approximation with 4 terms is shown in Figure 6.1 with the graph of sin(x) and the point with function value $\frac{32\pi}{180}$ indicated.

Historical contexts and applications

Examples of historical contexts pertinent to this Situation may be of interest to participants. For example, reference to Ptolemy's table of chords can be found in Maor (1998, pp. 25–28), and reference to the Rhind Papyrus, Problems 56–60, dealing with the Egyptian pyramids can be found in Maor (1998, pp. 6–9).

Thomas Paine (1737–1809), in the *Age of Reason* wrote:

The scientific principles that man employs to obtain the foreknowledge of an eclipse, or of anything else relating to the motion of the heavenly bodies, are contained chiefly in that part of science that is called trigonometry, or the properties of a triangle, which, when applied to the study of the heavenly bodies, is called astronomy; when applied to direct the course of a ship on the ocean, it is called navigation; when applied to the construction of figures drawn by a ruler and compass, it is called geometry; when applied to the construction of plans of edifices, it is called architecture; when applied to the measurement of any portion of the surface of the earth, it is called land-surveying. In fine, it is the soul of science. It is an eternal truth: it contains the mathematical demonstration of which man speaks, and the extent of its uses are unknown. (Paine, 1794, p. 32)

References to applications can be found in an NCTM publication (National Council of Teachers of Mathematics and the Mathematical Association of America, 1980, pp. 207–247); in Grazda, Brenner, and Minrath (1966); in Graham and Sorgenfrey (1989); in McConnell, Brown, Karafiol, Brouwer, and Ives (2016); in Peressini et al. (2016); and in Crisler and Froelich (2016), as well as in many other textbooks in trigonometry.

Reflect and assess learning

The purpose of reflection is for participants to connect the session's activities to their own classroom practice and to assess what participants learned from the session. Specific activities depend on the setting for the professional learning session, namely, whether it is a one-time session or part of an ongoing series of sessions.

Time

40–50 minutes

Suggested assessment activities

List three ideas that have extended your knowledge of the approximation of trigonometric ratios.

List three ideas you want to develop to explore with your students.

Was this session useful? What activities would you suggest to improve it?

Identify the Standards for Mathematical Practice with which you were engaged during the session.

Reflection questions

1. Has our work with this Situation caused you to consider or reconsider any aspects of your own thinking and/or practice about approximating trigonometric values? Explain.

2. Has our work with this Situation caused you to reconsider any aspects of your students' mathematical learning about approximating trigonometric values? Explain.

3. What additional questions has our work with this Situation raised for you?

NOTES

1. The Calculation of Sine Situation is one of the Situations presented in *Mathematical Understanding for Secondary Teaching: A Framework and Classroom-Based Situations* (Heid & Wilson, 2015).

2. This Situation appears on pp. 377–384 of Heid and Wilson (2015). It is reprinted with permission.

3. In the Common Core State Standards for Mathematics, additional mathematics that students should learn in order to take advanced courses such as calculus, advanced statistics, or discrete mathematics is indicated by (+).

4. See the Appendix for the list of the Standards for Mathematical Practice in the Common Core State Standards for Mathematics (National Governors Association Center for Best Practices and Council of Chief State School Officers, 2010).

REFERENCES

Crisler, N., & Froelich, G. (2016). *Precalculus: Modeling our world* (2nd ed.). Bedford, MA: COMAP.

Gelfand, I. M., & Saul, M. (2001). *Trigonometry*. Boston, MA: Birkhäuser Boston.

Graham, J. A., & Sorgenfrey, R. H. (1989). *Trigonometry with applications*. Boston: McDougal Littell/Houghton Mifflin.

Grazda, E. E., Brenner, M., & Minrath, W. R. (1966). *Handbook of applied mathematics* (4th ed.). Princeton, NJ: D. Van Nostrand.

Heid, M. K., & Wilson, P. W. (with G. W. Blume). (Eds.). (2015). *Mathematical understanding for secondary teaching: A framework and classroom-based situations*. Charlotte, NC: Information Age.

Maor, E. (1998). *Trigonometric delights*. Princeton, NJ: Princeton University Press.

McConnell, J., Brown, S., Karafiol, P., Brouwer, S., & Ives, M. (2016). *Functions, statistics, and trigonometry*. Chicago, IL: The University of Chicago.

National Council of Teachers of Mathematics and the Mathematical Association of America. (1980). *A sourcebook of applications of school mathematics*. Reston, VA: NCTM.

National Governors Association Center for Best Practices & Council of Chief State School Officers. (2010). *Common core state standards for mathematics*. Washington, DC: Authors. Retrieved from www.corestandards.org/uploads/Math_Standards1.pdf

Paine, T. (1794). *The age of reason*. Boston: Hall.

Peressini, A., DeCraene, P., Rockstroh, M., Viktora, S., Canfield, W., Wiltjer, M., & Usiskin, Z. (2016). *Precalculus and discrete mathematics*. Chicago, IL: The University of Chicago.

Usiskin, Z., Peressini, A., Marchisotto, E. A., & Stanley, D. (2003). *Mathematics for high school teachers: An advanced perspective*. Upper Saddle River, NJ: Pearson.

CHAPTER 7

FACILITATOR'S GUIDE
FOR
MEAN AND MEDIAN

Situation 38 From the MACMTL–CPTM Situations Project[1]

Rose Mary Zbiek with M. Suzanne Mitchell

The Situation selected for this chapter is one that is of new and increasing importance in school mathematics. As statistical content becomes a staple in school mathematics, teachers have an increasing need to develop a robust and flexible understanding of basic statistical concepts. They need not only to be able to calculate common measures of center and spread but also to be able to analyze relationships among those constructs. Essential to understanding those relationships is the ability to interpret representations of data sets. The pages that follow offer ideas for exploring, in either a professional learning or teacher preparation context, a Situation about mean and median in the context of common representations of data sets. The setting for the Prompt in this Situation was an Advanced Placement Statistics class, but the content of the Situation, particularly box plots and

Facilitator's Guidebook for Use of Mathematics Situations in Professional Learning,
pages 135–156.

comparison of data sets, is relevant to statistics topics in Grades 6–12. Box plots are part of middle school statistics content, and comparison of data sets foreshadows inference.

OVERVIEW

Facilitators can scan the following to quickly get a sense of the mathematics involved in the proposed professional learning setting. The sections of the overview deal with the mathematics of the Mean and Median Situation, why these mathematical ideas might be important for participants, the learning goals for participants, and mathematical ideas central to the proposed professional learning sessions.

Situation	Relevance
Given particular representations of data sets, exact analysis and precise conclusions about measures of central tendency such as mean and median may or may not be possible. This Situation addresses the use of mean and median to compare two samples given only box plots and five-number summaries.	• Box plots and five-number summaries are topics in middle school and high school mathematics. • Box plots and five-number summaries provide insights into distributions of quantitative data. They are useful tools in comparisons between data sets and can be used to foreshadow inference. • Observations about what can and cannot be read from box plots and five-number summaries have parallels with other representations of data, and with graphical or pictorial representations of mathematical and statistical objects other than data and distributions.
Goals	Key Mathematical Topics
• Participants will increase their understanding of box plots and five-number summaries and what each representation tells about data sets. • Participants will investigate relationships between mean and median when various representations of data sets are given. • Participants will learn an informal way to compare two sets of quantitative data.	• Information about means and quartiles that arise from box plots and five-number summaries • Statistics that are members of the data set— the minimum and maximum of a data set—and statistics that are not necessarily members of the data set—the median, first quartile, and third quartile • A box plot or a five-number summary as a representation of (finitely) many data points • Comparison of two samples by comparing measures of center and measures of variation

The Situation under consideration in this chapter follows (highlighted with a gray background).[2]

MEAN AND MEDIAN

Situation 38 From the MACMTL–CPTM Situations Project

Susan Peters, Evan McClintock, Donna Kinol, Shiv Karunakaran,
Rose Mary Zbiek, M. Kathleen Heid,
Laura Singletary, and Sarah Donaldson

PROMPT

The following task was given to students at the end of the year in an AP Statistics class.

Consider the box plots and five-number summaries[1] for two distributions, each of which is comprised of a finite number of data values (see Figure 44.1 and Figure 44.2). Which of the distributions (Data Set 1 or Data Set 2) has the greater mean?

One student's approach to this problem was to construct what he thought were probability distributions for each data set and to compare the corresponding expected values to determine which data set had the greater mean. The student formed four intervals using the five-number summaries and calculated the midpoint of each interval (i.e., he defined the intervals as the four quarters of the distributions, with each quarter containing 25% of the values for the distribution). Using the midpoint of each interval as the X-value for that interval, he then calculated the weighted mean for each probability distribution (see Figure 44.3). After completing his calculations, the student responded that the second data set had the larger mean.

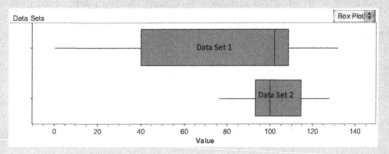

FIGURE 44.1. Box plots for Data Set 1 and Data Set 2.

Data Sets

	Group	
	one	two
Value	0	76
	40	93
	102	100
	109	115
	132	128

S1 = min ( )
S2 = Q1 ( )
S3 = median ( )
S4 = Q3 ( )
S5 = max ( )

FIGURE 44.2.   Five-number summaries for Data Set 1 and Data Set 2.

Data Set 1:

$E(X) = 79.25$

	0 – 40	40 - 102	102 - 109	109 - 132
X	20	71	105.5	120.5
P(X)	0.25	0.25	0.25	0.25

Data Set 2:

$E(X) = 102.75$

	76-93	93-100	100-115	115-128
X	84.5	96.5	107.5	121.5
P(X)	0.25	0.25	0.25	0.25

FIGURE 44.3.   The student's calculations for the two data sets.

COMMENTARY

This Prompt deals with the differences and similarities between the mean and median of a particular data set when the data set is displayed as a box plot using its five-number summary. It is likely that the intent of the question was not to encourage mathematical calculations, but rather to ask students to predict which distribution would have a greater mean based on what is expected from the visual display of the box plots.

Another important aspect to consider is that the problem given in the Prompt is given without a context. Without knowing the context of the data given, one does not know whether the data are continuous or discrete. Also, one does not know the statistical question being considered, namely, why the data were collected.

MATHEMATICAL FOCI

Mathematical Focus 1

The skewness of a data distribution affects the relationship between the mean and median of that set of data.

The box plots in Figure 44.1 display information about the distributions of the two data sets. Data Set 2 appears to be roughly symmetric with possible right skewness (note that the median is pulled to the left of central box). For this case, one would expect the mean and the median to be approximately equal, or the mean to be slightly greater than the median. Data Set 1 appears skewed left. If the distribution for the first data set is skewed to the left, then the smaller values have a stronger impact on the mean than the larger values. Therefore, one expects the mean to be less than the median. On the other hand, the median is the "middle" value of the data set after the data set is arranged in increasing order, and it is resistant to the larger spread in the smaller values. Because the medians are similar in each distribution, one expects Data Set 2 to have the larger mean. Although reasoning via the shape of the distribution is an approach that typically works when making comparisons about distributions, it is not always possible to make definitive statements about the relative locations of the means for some pairs of box plots.

Mathematical Focus 2

A box plot display of data does not necessarily give the data values or information about the "distribution" of the data within each quarter.[2]

How the data are distributed within each interval determined by the five-number summary is not represented in a box plot. The mean of a particular interval represented by the midpoint would be representative of the data points in that interval only when the data are distributed normally, uniformly, or symmetrically within that interval. The information given in the Prompt does not allow one to make such an assumption.

In the Prompt the student assumed that each interval contains exactly 25% of the data points. However, this is true only when the number of data points is divisible by 4. Also, some of the numbers in the five-number summary may not be members of the data set. The only values in the data set that are known for certain from the box plot are the minimum and maximum values. The median will be a member of the data set only when the number of data points is odd.[3] Q_1 and Q_3 are members of the data set only when the size of the data set has a remainder of 2 or 3 when divided by 4.[4]

Mathematical Focus 3

When exact values of two data sets are not known, comparisons between the two data sets can sometimes be made by comparing the ranges of their possible values.

For the given box plots, one can calculate an upper bound and a lower bound for the means of the data sets. Because of the apparent skewness discussed in Focus 1, one can assume that Data Set 2 has a greater mean than Data Set 1. To investigate this, consider the upper bound for the mean of Data Set 1 and compare it to the lower bound for the mean of Data Set 2. The greatest possible value of the mean of Data Set 1 is strictly less than the least possible value of the mean of Data Set 2. Therefore, it can be concluded that the mean of Data Set 2 is greater than the mean of Data Set 1.

To find an upper bound for the mean of Data Set 1, assume that in each interval the data points are located at the greatest possible value within the interval. (Of course, strictly speaking, since 0 is the minimum value of the data set, 0 is also a data point in the first interval; however, because an upper bound is being investigated and n is not known, 0 is not included in the calculation.) Let 25% of the values in Data Set 1 be located at the maximum value of each interval, that is, at Q1, at the median, at Q3, and at the maximum. In this extreme case, the mean of the data set is given by:

$$E(\text{Data Set 1}) = 0.25(40) + 0.25(102) + 0.25(109) + 0.25(132) = 95.75$$

Similarly, to find a lower bound of the mean for Data Set 2, assume that in each interval the data points are located at the least possible value within the interval. (Using this method, do not include the maximum value, 128, in the calculation.) Let 25% of the values in Data Set 2 be located at the minimum value of each interval, that is, at the minimum, at Q1, at the median, and at Q3. In this extreme case, the mean of the data set is given by:

$$E(\text{Data Set 2}) = 0.25(76) + 0.25(93) + 0.25(100) + 0.25(115) = 96.00$$

Because the lower bound of the mean of Data Set 2 is greater than the upper bound of the mean of Data Set 1, one can conclude that the mean of Data Set 2 is greater than the mean of Data Set 1.

The previous example assumed that each interval contained exactly 25% of the data set. This assumption may often be wrong, because this situation occurs only when the size of the data set is equivalent to $4n$ (i.e., is divisible by 4). The Post-commentary contains further investigation of Data Set 1, taking into consideration the possibility that the data set is of size

$4n + 1$, $4n + 2$, or $4n + 3$. Even in these cases, however, the mean of Data Set 2 can be shown to be greater than the mean of Data Set 1 (see Appendix to Focus 3).

Mathematical Focus 4

Stating a definitive conclusion about a comparison of the means using the five-number summaries and box plots is not always possible because the size of the data set may influence the relationship between the means for these distributions.

The following example portrays the importance of sample size and its effect on the relationships between the means and the five-number summaries for the given distributions.

For this example, consider the possibility that each data set contained 12 values (the box plots and five-number summary of which appear in Figure 44.4 and Figure 44.5). If the values in Data Set A were 14, 14, 17, 17, 17, 21, 21, 21, 24, 24, 24, and 30, then the mean of Data Set A would be 20.333. If Data Set B contained the values of 0, 12, 12, 12, 22, 22, 22, 26, 26, 26, 31, and 31, then the mean of Data Set B would be 20.167. Thus, the mean of Data Set A would be larger than the mean of Data Set B.

However, what if each set of data contained 100 values and the five-number summaries were maintained? If Data Set A were distributed with 24 values at 14, 25 values at 17, 25 values at 21, 25 values at 24, and 1 value at 30, then the mean of Data Set A would be 19.16. If Data Set B were distributed with 24 values at 31, 25 values at 26, 25 values at 22, 25 values at 12, and 1 value at 0, the mean of Data Set B would be 22.44. Thus, the mean of Data Set B would be larger than the mean of Data Set A.

Because the sizes of the data sets may influence the relationship between their means, stating definitive conclusions about a comparison of means is not always possible for a pair of box plots.

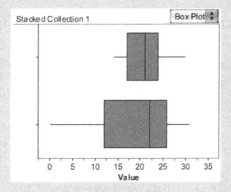

FIGURE 44.4.   Box plot for Data Set A and Data Set B.

	Group	
	One	Two
	14	0
	17	12
Value	21	22
	24	26
	30	31

S1 = min( )
S2 = Q1( )
S3 = median( )
S4 = Q3( )
S5 = max( )

FIGURE 44.5. Five-number summary for Data Set A and Data Set B.

POSTCOMMENTARY

Each of the Foci highlight a difference between information that allows conclusions to be made with mathematical precision and information that allows only for general claims to be made. For example, Focus 3 contains a conclusive argument for the relative sizes of the two means, whereas Focus 1 describes how claims can be made based on what is expected using reasoning about the shape of the distribution.

APPENDIX TO MATHEMATICAL FOCUS 3

In Focus 3, the upper bound of the mean of Data Set 1 and the lower bound of the mean of Data Set 2 were calculated assuming the size of each data set was divisible by 4. Figures 44.6, 44.7, and 44.8 illustrate the possibilities not considered in Focus 3 (i.e., that the size of a data set might be $4n + 1$, $4n + 2$, or $4n + 3$).

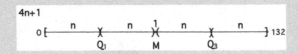

FIGURE 44.6. Data Set 1 contains $4n + 1$ data values.

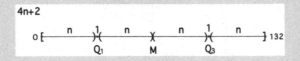

FIGURE 44.7. Data Set 1 contains $4n + 2$ data values.

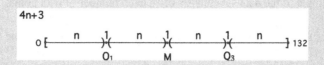

FIGURE 44.8.   Data Set 1 contains $4n + 3$ data values.

Consider the possibility that Data Set 1 contains $4n + 1$ data values. To maintain the same five-number summary, the extra

data value would be located at the median, 102. The mean value of 95.75 previously calculated for Data Set 1 assumed 25%

of the data values would lie in each segment; however, the lower extreme value of 0 was not accounted for in the calculation.

Note that accounting for the lower extreme of 0 (by effectively removing a value of 40) will lower the mean more than the

addition of a value at 102 will increase the mean. Thus, the mean of Data Set 2 is still greater than the mean of Data Set 1.

If Data Set 1 contains $4n + 2$ values, then one additional value will be located at Q1 and one additional value will be

located at Q3. Accounting for the minimum of 0, this situation nets the addition of a value at 0 and a value at 109. The value

added at 0 lowers the mean more than the value of 109 increases the mean, and the mean of Data Set 2 is still greater than

the mean of Data Set 1.

Finally, if Data Set 1 contains $4n + 3$ values, then a value is added at each of Q1, the median, and Q3. Accounting for the

minimum of 0, this situation nets the addition of a value at 0, a value at 102, and a value at 109. The value added at 0 lowers

the mean more than the addition of values both at 102 and 109. Therefore, the mean of Data Set 2 is still greater than the

mean of Data Set 1.

In all cases, the mean of Data Set 2 is greater than the mean of Data Set 1. Similar arguments can be made for different

sample sizes and their effects on the mean of Data Set 2.

NOTES

1.   Box plots are sometimes referred to as boxplots or box-and-whisker plots. The box plot is a visual display of the

five statistics values that comprise the five-number summary.

2.   A *quartile* is the boundary point for quarters of the data.

3.   Note that if there are multiple data values that all have the same value as the median, the median may not be a

specific member of the data set but will have the same value as members of the data set. In that case, the average of

the middlemost two values would equal the two values and hence the median would be equal in value to members of the data set. For example, if the data set consists of an even number of data points, say 5, 5, 5, and 5, the median is 5, which is not one of the specific members of the data set but which has the same value as each of the members of the data set.

4. The first and third quartiles were calculated using one particular convention. As noted by Kadar and Jacobbe (2013), "There are several different methods for determining quartiles. For example, when *n* is odd, the ordered data cannot be evenly divided in half. One method excludes the median from the lower and upper 'halves' when determining the quartiles, while another method includes the median in both 'halves'" (p. 35). The convention assumed in this Situation excluded the median from the lower and upper "halves," which is a method typically used in schools.

REFERENCE

Kadar, G., & Jacobbe, T. (2013). *Developing essential understanding of statistics for teaching mathematics in Grades 6–8.* Essential understanding series. (P. Wilson, Vol. Ed.; R. M. Zbiek, Series Ed.). Reston, VA: National Council of Teachers of Mathematics.

CONNECTION TO STANDARDS

Learning experiences in mathematics for teachers are well positioned when teachers are aware of connections to the standards for which their students are accountable. These standards differ across states and provinces as well as across countries, although there are commonalities across different sets of standards. Although this document cannot possibly address all the existing sets of standards across states, provinces, and countries, an example follows that illustrates how the proposed professional learning might address one particular set of standards. The example addresses the connections of the Common Core State Standards for Mathematics (CCSSM) (National Governors Association Center for Best Practices & Council of Chief State School Officers, 2010) to the proposed professional learning related to the Mean and Median Situation. The Common Core State Standards for mathematical content and mathematical practice mirror the attention given in various sets of standards both to what mathematics should be learned and to the mathematical processes in which students should engage. The Appendix of this Guidebook lists the CCSSM Standards for Mathematical Practice, one example of mathematical process standards. Questions (in bold italics in the following chart) that accompany the display of each CCSSM standard can be used in professional learning settings to extend work with specific standards.

Related Common Core Standards
CCSSM Standards for Mathematical Content
Grade 6 Statistics and Probability   **Develop understanding of statistical variability.**   6.SP.3. Recognize that a measure of center for a numerical data set summarizes all of its values with a single number, while a measure of variation describes how its values vary with a single number.   ***How do you use measures of variation when comparing measures of center?***[3]   6.SP.4. Display numerical data in plots on a number line, including dot plots, histograms, and box plots.   ***What can we tell about a data set when we look at a box plot for the data set?***   **Grade 7 Statistics and Probability**   **Draw informal comparative inferences about two populations.**   7.SP.4. Use measures of center and measures of variability for numerical data from random samples to draw informal comparative inferences about two populations.   ***How different do the box plots for two sample distributions have to be before we can reasonably conclude the populations are substantially different?***   **High School—Statistics and Probability**   **Summarize, represent, and interpret data on a single count or measurement variable.**   S-ID.1. Represent data with plots on the real number line (dot plots, histograms, and box plots).   ***What information about a data set does a box plot not convey?***
CCSSM Standards for Mathematical Practice[4]
SMP2. Reason abstractly and quantitatively.   **SMP3.** Construct viable arguments and critique the reasoning of others.   **SMP6.** Attend to precision.

SUGGESTIONS FOR USING THIS SITUATION

Facilitators may want to peruse what follows for an idea about how to use the Mean and Median Situation in their professional learning settings. The chart provides an Outline of Participant Activities and a summary of the Tools and projected Time required for implementation of those activities. Following the outline are Facilitator Notes that describe each of the suggested activities in greater detail.

Tools	Time
• Dynamic statistics environments or graphing calculators and/or other computing technologies that produce box plots and five-number summaries • Poster paper, markers (or a shared electronic document) • Copies of the Prompt (separate from the Foci) • Copies of the Foci	1–2 hours, in a single session or across multiple sessions
Outline of Participant Activities (Details for these activities follow in the Facilitator Notes section.)	
Launch Participants complete a chart indicating what two types of data displays—box plot and five-number summary—do and do not reveal about the data.	
Activity 1. After reading the Prompt, participants agree or disagree with the student's conclusion and explanation.	
Activity 2. The participants complete the original task as it was presented to the student and/or generate pairs of sets of values for the two data sets.	
Activity 3. Participants compare their approaches to the ideas in the Foci.	
Reflect and assess learning Participants revise their charts describing what can be concluded from box plots and five-number summaries. They also revise the Prompt's task to make it appropriate for middle school students, developing and critiquing lessons based on the revised task. Finally, they reflect on how their work with the Situation has affected their understanding of statistics.	

FACILITATOR NOTES

About the mathematics

There are multiple interpretations and misconceptions of *mean*. Two primary interpretations of mean are: mean as a leveler of data and mean as a balance point of a data set. Frequently, interpretations of mean emphasize calculations using a standard algorithm and ignore the conceptual understanding of the term or its properties. The following characteristics from O'Dell (2012) are ideas that students should develop, and which could be highlighted in discussions of this Situation:

- The mean is a value located between the maximum and minimum values of the data set.

- The mean does not have to be a member of the data set.

- For any data set, the sum of the signed distances of numbers from the mean of the data set is 0. (p. 149)

Mean and *median* are differently affected by the specific data values in a data set. Data Set 1 and Data Set 2 with the same five-number summaries (and therefore the same box plots) have the same medians. However, the mean of Data Set 1 might be greater than, equal to, or less than the mean of Data Set 2. Given a five-number summary and attempting to generate a pair of data sets for each of the three conditions illuminates how specific data values affect mean and median differently.

A *statistic* is "a numeric measure calculated from sample data, such as the mean or the standard deviation" (Peck, Gould, & Miller, 2012, p. 112). Mean, median, maximum, minimum, and quartiles are statistics—there might or might not be a data value equal to a mean, median, first quartile, or third quartile.

Statistics involves context. The task as presented to the student is devoid of context. As a result, we do not know whether the data are continuous or discrete. We also do not know how many data points are in the data set or what the data values are. Although it is tempting to create hypothetical data sets when reasoning about the Prompt, such data sets might be impossible in particular contexts.

Launch

In the launch, focus on the question of what particular types of data displays—namely, box plot and five-number summaries in this Situation—do and do not reveal about the data.

The Related Common Core Standards table for the Mean and Median Situation displayed earlier in this Guide identifies where ideas about box plots, five-number summaries, and means like those in this Situation occur in the Common Core State Standards for Mathematics.

Ask participants to complete the following chart based on experiences prior to this session:

	Box plot	Five-number summary
What can we tell with certainty about a data set by looking at this display?		
What can we approximate about a data set by looking at this display?		
What can't we tell about a data set by looking at this display?		

Time

3–5 minutes

Anticipated participant responses

In the chart that follows, the unbracketed entries are possible correct responses, and the entries in brackets are possible

flawed responses that should be challenged during this session.

	Box plot	Five-number summary
What can we tell easily and with certainty about a data set by looking at this display?	How, if at all, the data set is skewed [Value of the mean]	Value of the median, quartiles, maximum, and minimum [Five values in the data set] [Value of the mean]
What can we approximate about a data set by looking at this display?	Value of the median, quartiles, maximum, and minimum [Value of the mean]	[Value of the mean]
What can't we tell about a data set by looking at this display?	Number of values in the data set Whether the data set follows a normal distribution (A bimodal distribution could have a symmetric box plot.) Values of the mean and standard deviation	More than two values in the data set Number of values in the data set Whether the data set follows a normal distribution Values of the mean and standard deviation

Facilitating the launch

The given displays provide information about the median but not information about the mean. To encourage partici-

pants to continue their work on the task, ask them to write statements about statistics that they have heard of but have not yet included in the chart. For example, participants might be asked to write statements in their charts to describe what conclusions about range, mode, or standard deviation can be made from each of the two types of displays.

Let participants discuss in their groups their work on completing the chart; however, do not have groups report out. Chances are at least some of the participants' responses will be incomplete or incorrect. The purpose is for participants to reflect on their current knowledge of box plots and five-number summaries. This is an opportunity to assess the statistics background of different groups.

Activity 1. Have participants read the Prompt and then have them agree or disagree with the student's conclusion and explanation.

Have participants read the Prompt and give their initial impressions of (a) whether the student's conclusion is valid and (b) the extent to which the student's reasoning is valid. Initial impressions are the goal here; a later activity will have groups delve into constructing solid statistical arguments beyond reacting to the one student's response.

Time

10–15 minutes

Anticipated participant responses

Participants may wonder how it makes sense to think about the data in terms of four intervals. They may wonder: whether the midpoint for the interval is an appropriate value to use to represent the interval, what statistic the student is trying to calculate for each of the two data sets, or whether there is a data set for which the student's method works.

Facilitating the activity

Ask participants not to work through the students' work but to generate a list of questions that they would ask about the reasoning in order to evaluate it (e.g., Does it make sense to use a mean to represent each interval?). The focus on questions should draw participants' attention to the key ideas.

Participants might prematurely come to agreement or disagreement with the student's conclusion. To draw attention to the interpretation of the box plots and five-number summaries, ask participants to identify the decisions and claims that the student is making about the representations. Their identification of decisions and claims might be further facilitated by asking them to note claims and decisions that the student makes about things that the student named or used: intervals, means for intervals, midpoints for intervals, weights, data values, and number of data values. This attention to claims seems useful both in evaluating student reasoning and in drawing attention to the structure of the argument.

Participants might focus on instructing the student. This tendency needs to be balanced with their investigation of valid explanations (e.g., use of better interval estimates as in Focus 3) and alternative explanations (e.g., use of skewness as in Focus 1). [Note: It might be desirable initially to give participants the Prompt without the student's response.]

The idea of comparing intervals is a viable one. The challenge lies in choosing appropriate values to represent the data within each interval. Participants might be asked to identify alternative ways to choose a representative value for each interval.

Activity 2a. Have groups of participants work on the original task as it was presented to the student in the Prompt.

Groups might choose to refine the student's approach or to develop a completely different approach. Alternative approaches across groups and within groups should be encouraged. Ask each group to produce a poster that conveys its reasoning and conclusion.

Time

20–30 minutes

Anticipated participant responses

An initial response might be to observe that the box plot for Data Set 2 is "farther to the right" and thus would have a greater mean than Data Set 1, which is not necessarily true. Some groups likely will generate sets of data values and use these examples in reasoning about which distribution has the greater mean.

Facilitating the activity

As groups work, watch for concepts and generalizations used in their reasoning that might later connect to one or more of the Foci. Here are some possibilities:

- Skewness of the data (Focus 1)
- Whether 25% of the data values fall in each of four intervals (Focus 2)
- Whether the median and first and third quartiles are values in the data set (Focus 2)
- Comparison of intervals by estimating upper and lower bounds (Focus 3)
- Even or odd number of values in the data set (Focus 3)
- Validity of claims for other box plots or five-number summaries (Focus 4)
- Data sets of unequal size (Focus 4)
- Divisibility of the number of values in the data set by 4 (Postcommentary)

Ensure that the conclusion is clearly stated on each poster.

The student attempted to compare an estimated upper bound of the mean for Data Set 1 with an estimated lower bound for the mean of Data Set 2. One way to extend this argument is to assume that an equal number of values is added to each interval with the goal of maximizing the mean of Data Set 1 and minimizing the mean of Data Set 2. With that goal in mind, any value added to Data Set 1 would have to be 40, 102, 109, or 132 for the first, second, third, and fourth intervals, respectively. For Data Set 2, the values added to the first, second, third, and fourth intervals would be 76, 93, 100, and 115, respectively. Suppose that this were done n times. Then, the mean of the extended Data Set 1 would be

$$\frac{((0+40+102+109+132)+n(40+102+109+132))}{4n+5} = \frac{383+383n}{4n+5}.$$

The limit of this value of this expression as n approaches infinity is $383/4 = 95.75$. For extended Data Set 2, the mean would be

$$\frac{((76+93+100+115+128)+n(76+93+100+115))}{4n+5} = \frac{512+384n}{4n+5},$$

the limit of which as n approaches infinity is $384/4 \approx 96$. Thus, the mean of extended Data Set 2 would be greater than the mean of extended Data Set 1.

Activity 2b. (alternative to Activity 2a or support for Activity 2a) Ask participants to generate pairs of sets of values for the two data sets.

Challenge participants to produce pairs of data-set values for Data Set 1 and Data Set 2 with the following characteristics. Each pair of data-set values should correspond to the five-number summaries and the box plots. Participants should be asked to create (a) one pair of data-set values with the mean of Data Set 1 greater than the mean of Data Set 2, (b) one pair of data-set values with the mean of Data Set 2 greater than the mean of Data Set 1, and (c) one pair of data-set values in which the means are equal. Use of dynamic statistics software might be helpful with computations as participants work on this task. [Note: This task might be a particularly powerful starting point for teachers with minimal experience in doing and teaching statistics.]

Time

10–20 minutes

Anticipated participant responses

Responses here can vary greatly in complexity. Some groups might simply use the five statistics as five data values. Participants might not add the same number of data values to each interval.

Facilitating the activity

The numbers of values in the two data sets can differ. A particularly interesting challenge for groups who progress quickly on this task is to produce a data set that matches the five-number summary but has fewer than five values in it, or to explain why such a data set does not exist. Here is the essence of an argument for why four such values cannot be found for Data Set 1:

The minimum and maximum, 0 and 132, must be included. So, the data set must have at least two numbers. However, 102 is not the midpoint of 0 and 132, and so Data Set 1 contains more than two values. If there were exactly three values in Data Set 1, the three values are the minimum, maximum, and the median. Because 102 is not the midpoint of 0 and 132, the data set contains more than three values. Suppose there are exactly four values: 0, *a*, *b*, and 132. Then, the midpoints of the intervals with 0 and *a*, *a* and *b*, and *b* and 132 would be 40, 102, and 109, respectively. So,

$$\frac{0+a}{2} = 40 \quad \text{or} \quad a = 80,$$

$$\frac{b+132}{2} = 109 \quad \text{or} \quad b = 86, \text{ and}$$

$$\frac{a+b}{2} = 102 \quad \text{but} \quad \frac{80+86}{2} = 83 \neq 102.$$

This contradiction means our assumption that four values were possible is not feasible. There must be at least five values in Data Set 1.

[Note: This task might be done between sessions.]

Activity 3. Have participants compare their approaches to the ideas in the Foci.

Given copies of the Foci, participants should look for similarities between their reasoning and the reasoning in the Foci in terms of concepts, facts, or generalizations used. This task engages teachers in thinking about what constitutes similar reasoning rather than simply equivalent conclusions. It also recognizes the fact that not all the Foci provide complete solutions to the task given to the student.

Time

15–30 minutes

Anticipated participant responses

Participants might have difficulty drawing conclusions about the data sets from the values that appear in the five-number summary or in the box plots. They also might have misconceptions that each of the numbers in these two representations must be one of the data-set values. Participants who are less familiar with skewness might have difficulty reasoning about the effect of skewness on the relationship between mean and median.

Facilitating the activity

Observations made during group work provide insights into which Foci connect to each group's reasoning.

Key points about the Foci

- Note that the Prompt begins with a task for AP Statistics students (i.e., the student is required to determine which distribution has the greater mean, that of Data Set 1 or that of Data Set 2) and provides one student's response to the task. The intention of the Prompt is for the teacher–reader to take on the students' task and decide which distribution has the greater mean. The Prompt might also suggest a related task for the teacher–reader: Determine whether the student's reasoning and conclusion are valid. The Foci provide several ways to think about the students' task; they do not explicitly address the validity of the student's response.

- The Commentary notes that the Prompt lacks context for the data and seems to encourage interpretation of box plots and five-number summaries rather than calculations.

- Focus 1 centers on the relationship between median and mean given the skewness of the data. The means are compared based on how the means are related to the medians. The mean would be nearly equal to the median but slightly larger than it for relatively symmetric distributions that are skewed slightly to the right. The mean for a data set that is skewed left would be less than the median of the data set. Although Data Set 1 is skewed right and Data Set 2 is skewed left with the mean of the former less than the mean of the latter, the comparison is inconclusive. We do not have sufficient information to say how much less or greater the means are than their given medians.

- Focus 2 addresses assumptions about data values and distributions that are tempting to make about a box plot. In particular, we do not know whether the quartile values are members of the data set, whether the number of data values in each interval is odd or even, and how the data are distributed within any one of the four intervals. The Prompt does not provide enough information to support or refute any of these assumptions.

- Focus 3 uses ranges of values rather than exact values to compare the two means. The strategy assumes that each interval contains 25% of the data and that each value in an interval of data is equal to only the upper bound or the lower bound of that interval. The assumption that each interval includes 25% of the values in the data set is challenged in the Postcommentary. [Note: The omission of 0 when finding an upper bound of the first quartile of Data Set 1 makes the calculation cleaner. With this assumption, the expected value of upper bound for the first interval is calculated as 0.25(40). If 0 were included in the n values in the interval, the expected value of the upper board for interval would be $0.25(0 + (n - 1)40)/n$, which is less than 0.25(40). Similarly, 128 is not considered in the calculation for Data Set 2.

- Focus 4 addresses the effect of the size of the data sets on the extent to which a comparison of their means can be definitive. Note that the box plots used for this Focus are not the box plots that appear in the Prompt.

- The Postcommentary extends Focus 3 by examining the number of values in the data set.

Key points for discussion

- After presentations of the match of participants' solutions with Foci ideas, ask participants to identify factors that affect comparisons between the means of two data sets—not necessarily those in the Prompt—given only box plots and five-number summaries for the data sets. For example, *What would have to be true if we can make a decisive conclusion about which data set has the greater mean*? Revisit and revise the charts made at the beginning of the activity.

- Engage participants in a reflective discussion about their learning from this activity:

  o Which Foci had you previously considered?

  o Which Foci had ideas that were new to you?

  o Which arguments constructed by groups or found in Foci were most compelling?

  o What new insights did you gain, if any, from using box plots or five-number summaries to compare data sets from these different Foci?

  o What confusion remains about possible relationships between the median and mean of two data sets or use of box plots or five-number summaries to compare data sets?

Reflect and assess learning

The purpose of reflection is for participants to connect the session's activities to their own classroom practice, and to assess what participants learned from the session. Specific activities depend on the setting for the professional learning session, that is, whether it is a stand-alone session or part of an ongoing series of sessions.

Time

120–150 minutes

Suggested assessment activities

- Ask participants to return to the chart created at the beginning of the session(s) and revise it.

- Ask participants to revise the initial task given to the student in the AP Statistics setting so that it would be appropriate for middle school students, perhaps with the goal of having middle school students attend to what we can and cannot tell from box plots or with the goal of observing that the order of medians for two data sets might or might not be the same as the order of the means for those data sets. If this session is part of an ongoing series of professional learning sessions, participants could be asked to try their tasks in their classrooms, then bring classroom artifacts to the next session, in the spirit of a lesson study or modified lesson study.

  One way to debrief and evaluate participants' lessons is to do a modified gallery walk, in which, as participants read and review each lesson, they write questions about the lesson on Post-It® notes. As the facilitator, you might want to provide "expert commentary" on the lessons in addition to participants' comments. Participants could be asked to revise their lessons based on the feedback they receive.

- Ask participants to generate similar tasks that encourage students to attend to what can and what cannot be determined from other types of data displays (e.g., histograms, circle graphs). Participants might work in grade-level groups, and make a poster of their lesson/task/explanation. If this session is part of an ongoing series of professional learning sessions, participants could be asked to try their lesson/task/explanation in their classroom, then bring classroom artifacts to the next session, in the spirit of a lesson study or modified lesson study.

  This activity could involve responding to a single writing prompt, or different prompts with various types of data displays could be given to different groups. The previously described options for debriefing could be used here.

- Also consider asking participants to reflect on which of the Standards for Mathematical Practice they were engaged in during the session.

Reflection questions

1. Has work with this Situation caused you to consider or reconsider any aspects of your own thinking and/or practice about interpreting box plots and other data displays? Explain.

2. Has work with this Situation caused you to reconsider any aspects of your students' statistical learning about interpreting box plots and other data displays? Explain.

3. Has work with this Situation engaged you in mathematical practices? Explain.

4. How might work with this Situation suggest ways to look at other representations in addition to displays of data sets, such as function graphs or geometric diagrams? [Note: Adjust the representations to match curriculum for the grades or courses taught by participants.]

5. What additional questions has work with this Situation raised for you?

NOTES

1. The Mean and Median Situation is one of the Situations presented in *Mathematical Understanding for Secondary Teaching: A Framework and Classroom-Based Situations* (Heid & Wilson, 2015).

2. This Situation appears on pp. 397–404 of Heid and Wilson (2015). It is reprinted with permission.

3. For each standard, we have suggested questions (those in italicized, boldface type) that might serve as follow-up questions to address the mathematics of the standard.

4. See the Appendix for the list of the Standards for Mathematical Practice in the Common Core State Standards for Mathematics (National Governors Association Center for Best Practices and Council of Chief State School Officers, 2010).

REFERENCES

Heid, M. K., & Wilson, P. W. (with G. W. Blume). (Eds.). (2015). *Mathematical understanding for secondary teaching: A framework and classroom-based situations*. Charlotte, NC: Information Age.

Kadar, G., & Jacobbe, T. (2013). *Developing essential understanding of statistics for teaching mathematics in Grades 6–8*. Essential understanding series. (P. Wilson, Vol. Ed.; R. M. Zbiek, Series Ed.). Reston, VA: National Council of Teachers of Mathematics.

National Governors Association Center for Best Practices & Council of Chief State School Officers. (2010). *Common core state standards for mathematics*. Washington, DC: Authors. Retrieved from www.corestandards.org/uploads/Math_Standards1.pdf

O'Dell, R. S. (2012). The mean as balance point. *Mathematics Teaching in the Middle School, 8*, 148–155.

Peck, R., Gould, R., & Miller, S. (2012). *Developing essential understanding of statistics for teaching mathematics in grades 9–12*. Essential understanding series. (P. Wilson, Vol. Ed.; R. M. Zbiek, Series Ed.). Reston, VA: National Council of Teachers of Mathematics.

CHAPTER 8

CONCLUDING THOUGHTS

Rose Mary Zbiek

The work of developing mathematical understanding for secondary mathematics teaching never ends. Mathematics is a rich and intriguing field in which new connections and insights are always possible. The mathematical understanding of both individual classroom teachers and those responsible for facilitating teachers' learning grows by engaging in learning and in reflecting on their learning as well as by facilitating learning activities for others. Situations are one tool for developing personal mathematics that can be useful in the work of teaching secondary mathematics. They are a resource that can be used in professional learning and in teacher preparation. Appropriately, the final pages of this book offer an opportunity to reflect on the professional learning strategies used in the six example Situation Guides articulated in chapters 2–7 of this Facilitator's Guidebook. The closing pages also form a venue for informing future efforts of readers and Situation authors alike.

COLLECTION OF SITUATIONS

Chapters 2–7 of this book present suggestions for facilitating professional learning through six Situations from the collection of Situations developed by mathematics educators at The Pennsylvania State University and the University of Georgia. The six

Facilitator's Guidebook for Use of Mathematics Situations in Professional Learning,
pages 157–163.

TABLE 8.1.   Situations Grouped (loosely) by Content Strands in the Context of Secondary School
Mathematics Content

Content strand[a]	Chapter number[b]	Situation title
Number and Operation	**7**	**Division Involving Zero**
	8	**Product of Two Negative Numbers**
	9	Cross Multiplication
	10	Summing the Natural Numbers
	11	Modular Arithmetic
	14	Properties of i and Other Complex Numbers
	15	Square Root of i
Algebra and Function	12	Absolute Value Equations and Inequalities
	13	Absolute Value in Complex Plane
	16	Exponent Rules
	17	Powers
	18	Zero Exponents
	19	Multiplying Monomials and Binomials
	20	Adding Square Roots
	21	Square Roots
	23	Zero-Product Property
	24	Simultaneous Equations
	25	Graphing Inequalities Containing Absolute Values
	26	Solving Quadratic Functions
	27	**Graphing Quadratic Functions**
	28	Connecting Factoring With the Quadratic Formula
	29	Perfect-Square Trinomials
	31	Translation of Functions
	32	Parametric Drawings
Geometry and Measurement	30	Temperature Conversion
	33	Locus of a Point on a Moving Segment
	34	Constructing a Tangent Line
	35	Faces of a Polyhedral Solid
	36	Area of Plane Figures
	37	Area of Sectors of a Circle
	38	Similarity
	39	Pythagorean Theorem
	40	**Circumscribing Polygons**
Data Analysis and Statistics	**44**	**Mean and Median**
	45	Representing Standard Deviation
	46	Sample Variance and Population Variance
	47	Least Squares Regression
Trigonometry	22	Inverse Trigonometric Functions
	41	**Calculation of Sine**
	42	Graphing Sin(2x)
	43	Trigonometric Identities
Calculus and Beyond	48	The Product Rule for Differentiation
	49	Proof by Mathematical Induction

Note: Boldface type indicates the Situations for which a guide appears in this Facilitator's Guidebook.
[a] The Situations typically involve mathematics that fits well in several categories.

[b] Chapter Number identifies the Situation's Chapter from Mathematical Understanding for Secondary Teaching: A Framework and Classroom-Based Situations (Heid & Wilson, 2015).

example Situations, and 37 more Situations, appear in *Mathematical Understanding for Secondary Teaching: A Framework and Classroom-Based Situations* (Heid & Wilson, 2015). In Table 8.1, the 43 Situations appearing in that book are grouped by content strands related to the context of secondary school mathematics.

INSIGHTS FROM EXAMPLES

Looking across the six Situation Guides offers several insights into how to structure and support professional learning around these and other contexts. Insights shared in this section are arranged under particular aspects of what a facilitator might anticipate or encounter while planning or facilitating a professional learning experience: Engaging With Situations, Isolating the Mathematics, Focusing on Mathematical Practices or Processes, and Extending the Learning. In the following sections, the discussion of each aspect includes insights into the nature of the mathematical learning that might be underscored or a sense of other learnings that complement the mathematics. Examples of particular moves or opportunities to develop these learnings include references to relevant components of the six Situation Guides in chapters 2–7 of this Guidebook.

Engaging with Situations

Fundamental to the use of Situations in professional learning is engagement with various parts of each Situation. A productive professional learning experience naturally provides opportunities and time for participants to invest in Prompts, to own the Foci, and to reflect on their experiences.

Investing in Prompts. Each of the Prompts in *Mathematical Understanding for Secondary Teaching* (Heid & Wilson, 2015) is derived from the work of teaching. Educators can benefit from taking time to read the Prompts and react to them based on their current understanding, before encountering the Foci and Commentaries. For example, facilitators might have participants read a Prompt and determine the correctness of its content (e.g., Activity 1 of Mean and Median, pp. 149–150). Participants might be encouraged to generate a variety of ways to look at a common misconception associated with the Prompt (e.g., Division Involving Zero, p. 9). For any Situation, participants might initially construct potential responses to the Prompt. Participants' responses could then be used to generate a collection of possible approaches and to connect their initial thinking to the content of particular Foci. Later, participants can return to their initial responses and revise them (e.g., Activity 1 and Activity 3 in Division Involving Zero, pp. 19–21; Activity 1 in Graphing Quadratic Functions, pp. 70–71; Activity 2 in Mean and Median, pp. 150–152; Activity 3 in Mean and Median, pp. 152–154).

Owning the Foci. The mathematics captured in a Focus for any one Situation can differ in complexity or depth from that in other Foci. Any participant likely will find at least one Focus to be challenging. Participants might attempt to understand the mathematics, the prerequisite understandings, and/or the assumptions in one of the Foci and present their analysis to the rest of the group (e.g., Activity 4 in Division Involving Zero, pp. 21–23; Activity 2 in Graphing Quadratic Functions, pp. 71–72; Activity 3 in Graphing Quadratic Functions, pp. 72–77).

Reflecting on experiences. Reflecting on experiences like one described in a Situation is as important as investing in the Prompts. Both actions can help to connect participants' work with a Situation, and newly developed mathematical understandings, to their classroom practice. Participants can identify ways in which the topic might arise in their classrooms and how they might foster learning related to that topic (e.g., Division Involving Zero, p. 19; Product of Two Negative Numbers, pp. 55–56; Graphing Quadratic Functions, p. 82).

After participating in an activity centered on a Situation, participants can tell how they might respond if they found themselves in the setting described in the Prompt. If participants recorded their initial thinking about the Prompt, as suggested previously, they can return to those ideas to apply their professional learning by revising those initial reactions (e.g., Mean and Median, pp. 154–155). They might, in addition or alternatively, share how their work on the Situation has changed their understanding of the mathematics (e.g., Division Involving Zero, p. 24; Product of Two Negative Numbers, p. 57; Mean and Median, p. 156).

Isolating the mathematics

Throughout engagement with Situations, opportunities arise for different kinds of fundamental mathematical points and practices or processes. Definitions, variation, and representations merit explicit mathematical attention.

Acknowledging definitions. Definitions are essential in mathematics as a field (Edwards & Ward, 2008; Leikin & Winicki-Landman, 2001; Zandieh & Rasmussen, 2010), and Situations offer opportunities to engage deeply with definitions. Participants can be asked not simply to draw on familiar descriptions of concepts but to recall definitions involved in key mathematical topics (e.g., Launch in Graphing Quadratic Functions, pp. 68–70).

Capturing variation. Acknowledging what does and does not vary is critical in mathematics (Sinclair, Pimm, & Skelin, 2012). Statements about what does not vary under particular conditions can describe observations in mathematical investigations and eventually be established as theorems. Participants can consider variations in the Prompt through, for example, changing parameters (e.g., Activity 4 in Graphing Quadratic Functions, p. 77).

Generalizing appropriately. Observations about invariance or about features revealed through examination and interpretation of mathematical representations are some examples of generalities in mathematics. Taking into consideration a statement and the domain of mathematical objects to which it applies is inherent in developing and using mathematical generalities (Gray-say, 2016). Students can implicitly act on or explicitly state generalities. In some cases, students state questionable properties or identify inappropriate domains. Participants might consider generalizations of relationships that are in the Prompt (Activity 5 in Graphing Quadratic Functions, pp. 77–81).

Representing effectively. Representations play a major role in how students understand concepts and definitions (e.g., Ainsworth, 2006), how they notice and convey invariance (e.g., Sinclair, Pimm, & Skelin, 2012), and how they encounter generality (e.g., Ellis, 2011). Curricular goals often draw attention to translating from one representation to another (e.g., National Governors Association Center for Best Practices & Council of Chief State School Officers, 2010). Situations offer opportunities for teachers to learn to value translating between representations and to think about and interpret potentially less familiar types of representations. For example, given a representation, participants might be asked to decide what can and cannot be concluded from that representation (Launch in Mean and Median, pp. 147–149).

Focusing on mathematical practices or processes

Beyond providing opportunities to work with definitions, invariants, representations, and generalizations, engagement with Situations leads to authentic experiences with what have been called *mathematical habits of mind* (Cuoco, Goldenberg, & Mark, 1996), *practices* (National Governors Association Center for Best Practices & Council of Chief State School Officers, 2010), or *processes* (Zbiek, Heid, & Blume, 2012). In particular, for example, participants could analyze arguments concerning the extent to which the reasoning is intuitive or deductive (e.g., Activities 2 and 3 in Product of Two Negative Numbers, pp. 54–56). More generally, participants could identify and debate which Mathematical Practices are addressed in each Focus (e.g., Activity 4 in Product of Two Negative Numbers, p. 56).

Extending the learning

The Situations offer opportunities to extend professional learning to areas other than mathematics content per se. Of particular relevance are digital technologies and curriculum.

Employing digital technologies. With the rise in technology available in schools (Herold, 2016; U.S. Department of Education, 2017), use of digital tools for learning and teaching is increasingly important. Learning to use tools for teaching mathematics is complicated, and substantial evidence suggests that secondary mathematics teachers can benefit from using technology when learning new mathematics (Strutchens et al., 2016). Even if the Prompt does not directly involve digital technologies, facilitators and teachers can consider whether digital technologies can afford some insight (e.g., Activity 2 in Division Involving Zero, pp. 19–20). Participants might use a technological tool to represent some aspect of what is in a Prompt (e.g., Launch in Graphing Quadratic Functions, pp. 68–70). Options might also exist for participants to have personal experiences using digital mathematics tools or collaboration/communication technology while they consult with peers and resources to unpack Prompts or work through Foci.

Connecting with curriculum and students. Many Situations, and especially the Foci, address content topics that arise in various ways in more than one grade level or school mathematics topic area. If participants identify with teaching at different grade levels, they could describe incidents similar to the Prompt that might occur in their settings and elaborate on the relevance of the Foci to their daily work. They could discuss where the topic in the Prompt appears in the mathematics curriculum for their grade or course (e.g., Launch in Division Involving Zero, p. 18; Launch in Product of Two Negative Numbers, pp. 46–49). Alternatively, participants could revise the task in the Prompt to address a different grade level (Mean and Median, p. 155). Attending to Foci, participants could consider how the content of the Foci connects to other topics in the secondary school curriculum (e.g., Activity 5 in Graphing Quadratic Functions, pp. 77–81). To connect the content with their awareness of students and classroom practice, they could rank arguments in the Foci or the mathematics in the Foci regarding how convincing or accessible they would be to students at various levels (e.g., Activity 1 in Product of Two Negative Numbers, pp. 49–53).

In summary, various facilitation strategies, such as those identified at the beginning of this section, can enhance teachers' experiences. Work with Situations offers many ways, not limited to those described in this section, in which teachers can enrich their understanding of mathematics for secondary teaching.

WELL WISHES

Teaching secondary mathematics calls for a special understanding of mathematics and facility in drawing on that understanding to respond to daily events that arise in the work of teaching. Situations are descriptive of mathematics related to incidents in the daily work of teaching. The purpose of this Guidebook is to engage teachers in professional learning grounded in such events. The suggested professional learning activities focus on teachers developing deeper understanding of school mathematics and its internal connections as well as its links to mathematics across grades and college courses. Readers are encouraged to see, explore, and be ready to use the mathematics that lies beneath, within, and beyond the Prompts and Foci. The chapters also provide suggestions and examples regarding how other Situations from *Mathematical Understanding for Secondary Teaching* (Heid & Wilson, 2015), as well as Prompts based on the reader's own experiences, can be used formally or informally by individuals or with groups to differently engage teachers in mathematics—and for all to enjoy doing so.

REFERENCES

Ainsworth, S. (2006). DeFT: A conceptual framework for considering learning with multiple representations. *Learning and Instruction, 16,* 183–198. doi:10.1016/j.learninstruc.2006.03.001

Cuoco, A., Goldenberg, E. P., & Mark, J. (1996). Habits of mind: An organizing principle for mathematics curricula. *Journal of Mathematical Behavior, 15,* 375–402. doi:10.1016/s0732-3123(96)90023-1

Edwards, B., & Ward, M. (2008). The role of mathematical definitions in mathematics and in undergraduate mathematics courses. In M. P. Carlson & C. Rasmussen (Eds.), *Making the connection: Research and teaching in undergraduate mathematics education.* Washington, DC: Mathematical Association of America.

Ellis, A. B. (2011). Generalizing-promoting actions: How classroom collaborations can support students' mathematical generalizations. *Journal for Research in Mathematics Education, 42,* 306–343. http://www.nctm.org/publications/jrme.aspx

Graysay, D. T. (2016). *Undergraduate students' approaches to constructing mathematical generalities* (Unpublished doctoral dissertation). The Pennsylvania State University, University Park, PA.

Heid, M. K., & Wilson, P. S. (with Blume, G. W.) (Eds.). (2015). *Mathematical understanding for secondary teaching: A framework and classroom-based situations,* Charlotte, NC: Information Age.

Herold, B. (2016, February 5). Technology in education: An overview. *Education Week.* Retrieved from http://www.edweek.org/ew/issues/technology-in-education

Leikin, R., & Winicki-Landman, G. (2001). Defining as a vehicle for professional development of secondary school mathematics teachers. *Mathematics Education Research Journal, 3,* 62–73.

National Governors Association Center for Best Practices & Council of Chief State School Officers. (2010). *Common core state standards for mathematics.* Washington, DC: Authors. Retrieved from http://www.corestandards.org/assets/CCSSI_Math%20Standards.pdf

Sinclair, N., Pimm, D., & Skelin, M. (2012). *Developing essential understanding of geometry for teaching mathematics in grades 9–12.* Essential Understanding Series (P. Wilson, Vol. Ed.; R. M. Zbiek, Series Ed.). Reston, VA: National Council of Teachers of Mathematics.

Strutchens, M. E., Huang, R., Losano, L., Potari, D., de Costa Trindade Cyrino, M. C., Pedro da Ponte, J. P., & Zbiek, R. M. (2016). *The mathematics education of prospective secondary teachers around the world* (ICME – 13 Topical Survey). Cham, Switzerland: Springer.

U.S. Department of Education. (2017). *Reimagining the role of technology in education: 2017 National Education Technology Plan Update.* Retrieved from https://tech.ed.gov/files/2017/01/NETP17.pdf

Zandieh, M., & Rasmussen, C. (2010). Defining as a mathematical activity: A framework for characterizing progress from informal to more formal ways of reasoning. *Journal of Mathematical Behavior, 29,* 57–75. doi:10.1016/j.jmathb.2010.01.001

Zbiek, R. M., Heid, M. K., & Blume, G. W. (July, 2012). *Seeing mathematics through processes and actions: Investigating teachers' mathematical knowledge and secondary school classroom opportunities for students.* Paper presented at the quadrennial meeting of the International Congress on Mathematical Education, Seoul, Republic of Korea.

APPENDIX: STANDARDS FOR MATHEMATICAL PRACTICE

Like many sets of curriculum standards, the Common Core State Standards for Mathematics (National Governors Association Center for Best Practices & Council of Chief State School Officers, 2010) blends both mathematics content and something more. For this something more, Common Core State Standards "rest on important 'processes and proficiencies' with longstanding importance in mathematics education" (National Governors Association Center for Best Practices & Council of Chief State School Officers, 2010, p. 6). In particular, the Common Core State Standards draws on the National Council of Teachers of Mathematics (NCTM, 2000) process standards and the strands of mathematical proficiency specified in the National Research Council's report *Adding It Up* (National Research Council, 2001). The following eight statements are the mathematical practices in the Common Core State Standards for Mathematics.

1. Make sense of problems and persevere in solving them.

2. Reason abstractly and quantitatively.

Facilitator's Guidebook for Use of Mathematics Situations in Professional Learning,
pages 165–166.
Copyright © 2018 by Information Age Publishing

3. Construct viable arguments and critique the reasoning of others.

4. Model with mathematics.

5. Use appropriate tools strategically.

6. Attend to precision.

7. Look for and make use of structure.

8. Look for and express regularity in repeated reasoning.

REFERENCES

National Council of Teachers of Mathematics. (2000). *Principles and standards for school mathematics*. Reston, VA: Author.

National Governors Association Center for Best Practices & Council of Chief State School Officers. (2010). *Common core state standards for mathematics*. Washington, DC: Authors. Retrieved from http://www.corestandards.org/wp-content/uploads/Math_Standards1.pdf

National Research Council. (2001). *Adding it up: Helping children learn mathematics* (J. Kilpatrick, J. Swafford, & B. Findell, Eds.). Washington, DC: National Academy Press.

Printed in the United States
By Bookmasters